Healing Architecture
治愈性建筑

［芬］埃萨·皮罗宁（Esa Piironen）著 ｜ 张亚萍 译

中国电力出版社
CHINA ELECTRIC POWER PRESS

内容提要

纵观历史，人类一直致力于通过建筑来为人类创造更好的环境。实现这一目标有许多方式，其中医院建筑是有助于人类健康的一个重要领域，但也有很多其他的建筑能够为我们的环境提供一种治愈性的体验。这一新兴的治愈性建筑，作为一种为人类建造更好建筑与更佳环境的方式，刚刚开始崭露头角。本书从堪舆学、医院设计、发展中的治愈性建筑、环境心理学、神经科学、治愈性建筑的设计元素（光、噪声、空气、自然、色彩、形式、尺度与比例、人体工程学、材料、氛围）等不同方面，论述了治愈性建筑作为一种建筑新范式的缘由、设计因素与未来发展。书中也指出在当下与未来，健康设计将成为建筑设计的新关注点之一，成为未来建筑设计的一个新方向，而治愈建筑将成为未来建筑设计的一种新范式。

图书在版编目（CIP）数据

治愈性建筑：汉英对照 /（芬）埃萨·皮罗宁（Esa Piironen）著；张亚萍译. —北京：中国电力出版社，2021.9

ISBN 978-7-5198-4699-2

Ⅰ.①治⋯ Ⅱ.①埃⋯ ②张⋯ Ⅲ.①建筑设计—汉、英 Ⅳ.① TU2

中国版本图书馆 CIP 数据核字（2020）第 099815 号

出版发行：中国电力出版社
地　　址：北京市东城区北京站西街 19 号（邮政编码 100005）
网　　址：http://www.cepp.sgcc.com.cn
责任编辑：王 倩（010–63412607）
责任校对：黄 蓓 李 楠
装帧设计：锋尚设计
责任印制：杨晓东
印　　刷：北京博海升彩色印刷有限公司
版　　次：2021 年 9 月第一版
印　　次：2021 年 9 月北京第一次印刷
开　　本：710 毫米 ×980 毫米　16 开本
印　　张：7.25
字　　数：122 千字
定　　价：68.00 元

建筑新范式

New Paradigm of Architecture

Foreword

Throughout history man has tried to create a better environment for people by architecture. There have been many ways to do that. Hospital building is one area that has helped people's wellbeing, but there are also other ways of building that has made our environment a healing experience. This new healing architecture is just emerging again as a mean to build better buildings and environment for human beings.

We build three types of housings: Sick buildings, where people become ill and are unpleasant in one way or another, buildings that are neutral, houses that do not hurt inhabitants and buildings that can heal people like home at its best.

History shows us that these so−called healing architecture could be a better solution for modern architecture. In history you can find examples to show that the healing architecture was discovered about five thousand years ago.

The text and images are mostly based on my lectures in China.

前言

纵观历史，人类一直致力于通过建筑来为人类创造更好的环境。实现这一目标有许多方式，其中医院建筑是有助于人类健康的一个重要领域，但也有很多其他的建筑能为我们的环境提供一种治愈性的体验。这一新兴的治愈性建筑作为一种为人类建造更好建筑与更佳环境的方式刚刚开始崭露头角。

我们建造的建筑有三种：病态的建筑，这些建筑使人们得病，用不同的途径令人感到不舒适；中性的建筑，对居住者来说是不好不坏的建筑；治愈性建筑，像最舒适的家一样能治愈人们的建筑。

历史告诉我们，所谓治愈性建筑可能是现代建筑一个更好的解决方案。历史表明，治愈性建筑在 5000 年前就已被发现。

本书中的文字与图片主要来自于我在中国的学术讲座系列。

Contents

目录

Introduction

"A House Can Hurt, a Home Can Heal" is a saying by an African–american author Maya Angelou, that I heard years ago. I started to think about healing architecture in a new way.

Modern architecture is becoming more diverse and a way more troublesome all over the world. Star architects design and build more and more strange and odd forms that are not based on the function of a building or try to harmonize with the environment built.

Star architects want to play with forms and try to shock people. Architecture is not an area to make experiments with people and their wellbeing.

Even the leaders in China are worried about what the foreign star architects have designed. Strange forms are governing the design and not the function or harmony as it should be.

An American architectural critic Charles Jencks published 2012 a small red book called *Can Architecture Affect Your Health*.The book was based on his lecture for Sikken–color firm. In the book he brings us information about ancient examples of healing environments that are not so well known.

很多年前我听过这样一句话："房子能伤人，家却能治愈人。"这句话出自非裔美籍作家玛雅·安杰罗（Maya Angelou），它让我开始用一种新的方式来思考治愈性建筑。

全世界的现代建筑正在变得越来越多样，越来越繁复。明星建筑师们设计和建造形式越来越稀奇古怪的建筑，这些建筑的形式并不是基于建筑的功能，也不试图与建造环境相协调。

明星建筑师们想要玩形式，并试图让人们感到震惊。但建筑行业不是一个用人们和他们的健康做实验的领域。

甚至中国的领导人也担心这些国外的明星建筑师们设计的建筑。奇怪的形式控制着设计，而不是以它应有的功能或和谐。

美国建筑评论家查尔斯·詹克斯（Charles Jencks）于 2012 年出版了一本小红书《建筑能影响你的健康吗》。这本书是基于他在锡肯色彩公司的演讲写成的。书中他带给我们一些不太为人知的古代治愈性环境的案例。

Archaeologists Geoffrey Wainwright and Timothy Darvill did research in Stonehenge UK in 2006. They showed that the place was a healing environment five thousand years ago. Bones have been found there show by DNA research that people even from Switzerland travelled there to get some cure. Maybe the other megaliths sites in Europe had the same purpose.

Epidauros in Greece is one of the first (400 BCE) places where healing was a way of life. There were spas, stadium, gyms, all kinds of temples and houses where the people who had come to heal stayed. Only the auditorium is still there, but the natural beauty and the view to the sea is outstanding and of course the fine acoustics.

图 1　Stonehenge 3000 BCE　巨石阵遗址，公元前 3000 年

2006 年，考古学家杰弗里·温赖特（Geoffrey Wainwright）和蒂莫西·达维尔（Timothy Darvill）对英国巨石阵做了研究（图 1）。他们的研究表明这里在 5000 年前是一个治愈性的环境。在此发现的人骨遗骸 DNA 研究表明，即使是来自瑞士的人也曾到此寻求治疗。可能欧洲的其他巨石阵也有同样的目的。

古希腊的埃皮道罗斯遗址（公元前 400 年）是最早将治疗作为一种生活方式的地方之一。这里有温泉浴场、体育场、健身房、各种寺庙和可供来治疗疗养的人们下榻的房间。如今遗址只剩剧场观众席，但是自然风光和海景依然非常美丽，当然还有美妙的音响效果（图 2）。

图 2　Epidauros 400 BCE　埃皮道罗斯遗址，公元前 400 年

Feng Shui

Feng Shui is a Chinese way to make healing architecture.

Feng Shui is an old eastern wisdom that tells us how forms, colors and materials affect us. It deals with the effects of visible and non–visible magnetic fields on us. It has been studied at least three thousand years. They say in the east that it is not a phlosophy, it is not a science nor a belief and not a religion.

The energies around us have certain formulas and we human beings should live in balance with these energies and our environment.

The most important ideas of Feng Shui are the five elements and jin and jang.The five elements are water, fire, wood, earth and metal. All the materials are based on these five elements. The elements contain energies that affect human beings in different ways. Yin and Yang are contrasts that work combined: light and darkness, strength and weakness, woman and man. If you know the qualities of these elements, you can arrange them in a harmonious way.

Like the differences in Western medicine and Chinese medicine, we should study to know better ways to heal human beings. We could learn from old traditions some good advice to build while we get new information of neuroscience for healing architecture.

堪舆学

堪舆学（也称风水）是源自中国的一种建造治愈性建筑的方式。

风水是一种古老的东方智慧，告诉我们建筑的形式、颜色和材料是如何影响我们的。风水能解决可见和不可见的磁场对我们产生的影响，至少已有 3000 年的研究历史。在东方，人们认为风水不是哲学，不是科学，不是信仰，也不是宗教。

我们周围的能量有一定的组合方式，因此人类必须与这些能量以及我们的生存环境之间保持一定的平衡。

风水最重要的观念是五行和阴阳。五行指的是金、木、水、火、土。所有的材料都基于这五行。五行包含着用不同方式影响人类的能量。阴阳相对，但结合在一起才有效：明与暗、强与弱、男与女。如果你了解这些元素的特性，则可以采用和谐的方式安排它们（图 3）。

如同西医和中医的差别，我们应该研究出更好的方式来治愈人类。当我们获取神经科学方面的新信息并用于建造治愈性建筑时，我们也应该从古老的传统中汲取好的建议。

The natural way of healing in chinese medicine is challenging, starting with acupuncture.

Healing architecture should take all natural aspects of design as a starting point for creating wellbeing for our inhabitants.

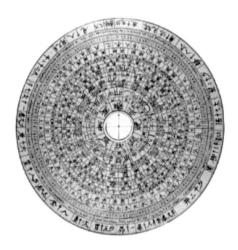

图 3　Feng Shui compass　风水罗盘

从针灸开始，中医的自然疗法很有挑战性。

治愈性建筑应当将考虑的所有自然因素作为设计出发点，为居住者创造幸福（图4）。

图4 Different ways to design with Feng Shui 利用风水原理的不同设计方式

Hospital Design

After these healing places hospitals came 400 years later. Monasteries were shelters for sick people, but nobody thought that buildings could heal people. Christian charity—the outcome of the council of Aix–la–Chapelle in 1816 introduced the Hotel–Dieu, the first hospital. In hospitals sick people were healed with care and medicine.

The earliest anatomic teaching place (theatre) was built in Padua 1594, where it was possible to analyse what was inside human beings. Afterwards one of the major healing process was through operations. Buildings were not thought to have any affect on sick people.

After the Hotel–Dieu was burnt in 1772 people were thinking that hospitals could be healing buildings. In 1879 Jean–Baptiste Leroy coined the term "machine à guérir": hospitals should be seen as healing machines.

The original concept of the healing environment was developed by Florence Nightingale whose theory of nursing called for nurses to manipulate the environment to be therapeutic. Nightingale outlined in detail the requirements of the "sickroom" to minimize suffering and optimize the capacity of a patient to recover, including quiet, warmth, clean air, light, and good diet. Early healthcare design followed her theories outlined in her treatise, *Notes on Hospitals*.

医院设计

在这些治愈性场所出现 400 年后，医院出现了（图 5）。修道院是病人
的庇护所，但是没人认为修道院建筑可以治愈人们。基督教慈善团体
1816 年在艾克斯拉沙佩勒会议上介绍了第一家医院——迪厄旅馆。在
医院里，生病的人们通过护理和药物来得到治愈。

1594 年，最早的教学地（解剖室）在意大利帕多瓦建成，这可能是人
们最早分析人体内部结构的建筑空间。之后，手术成为一种主要的治
疗手段。当时人们并不认为建筑对病人会产生任何影响。

1772 年迪厄旅馆被烧毁后，人们开始思考医院可以是治愈性的建筑空
间。1879 年，让-巴伯蒂斯特·勒罗伊（Jean-Baptiste Leroy）创造了"机
器般的治疗"一词：医院应该被视为一个治疗"机器"。

治愈性环境的最初概念是由弗洛伦斯·南丁格尔（Florence Nightingale）
提出的，她的护理理论要求护士布置好环境，以达到治疗的目的。南丁
格尔仔细列出了能最大限度地减少病人痛苦并帮助病人康复的"病房"
要求，包括安静、温暖、空气清新、光线适宜和良好的饮食。早期的医

Following the discoveries by Louis Pasteur and other which lead to the Germ Theory, plus other technologies, the role of the environment was dominated by infection control and technological advances.

Florence Nightingale, as long as 1859, identified key links between the physical design of hospitals and the very high level of mortality or prolonged illness of patients at that time. She even said that "The object and color in the materials around us actually have a physical effect on us, on how we feel".

The "medicalization" of the hospital began in the mid-19th century with reintegrigation of surgery in the medical world. Surgery required specific facilities, designating operating theatres as the first spaces that could only be found in hospitals. New spaces were needed when X-ray machines went to hospitals in 1897. Most of the thinking was concentrated to machinery, not to the healing building design.

Functionalism tried to make a design language for a healthy and aesthetic environment, but as we know, it forgot a human being, environment and history.

Alvar Aalto used psychology as a starting point of design when he designed Paimio TB sanatorium in Finland 1928–1933.

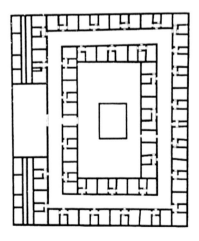

图 5 Roman military hospital, first century A.D. 罗马军事医院，公元 1 世纪

疗护理空间设计遵循了她在论文《医院笔记》中概述的理论。路易斯·巴斯德（Louis Pasteur）等人的发现带来了微生物理论的诞生，加之其他技术，感染控制和技术进步主导了环境的作用。

早在 1859 年，弗洛伦斯·南丁格尔就发现了医院建筑设计与当时非常高的死亡率或者病人久治不愈之间的关键关系。她甚至说："我们周围材料的对象和颜色，实际上对我们的感受有实质性的影响。"（图 6）

19 世纪中叶，随着外科手术在医学界的重新整合，医院开始医学化。外科手术需要特定的设施，作为首要空间的特定的手术室只有在医院才能找到。1897 年，当 X 射线机出现时，新的空间需求随之出现。关于医院设计的思考更多与医疗器械的放置有关，而不是建筑本身的治愈性设计。

功能主义试图创造一种健康、美观的环境设计语言，但如我们所知，它忘记了人类、环境和历史。

1928–1933 年，阿尔瓦·阿尔托在设计芬兰帕米奥（Paimio）结核病疗养院时，以心理学作为设计的出发点（图 7）。

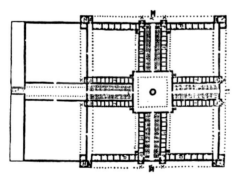 

图 6 Monastery plan, late Middle Ages 修道院建筑平面，中世纪晚期

图 7 Paimio Sanatorium, Alvar Aalto,1928-1933 帕米奥结核病疗养院，阿尔瓦·阿尔托，1928-1933

That building is one of the first in Finland to show new functionalism features in architecture. International trends came to Finland quite soon, because Alvar Aalto attended international CIAM conferences in Europe. He learned new ideas in architecture through Walter Gropius and especially Le Corbusier.

Alvar Aalto listened to the advices of the doctors involved in the design of a new sanatorium in Paimio. So the empathy was also known by Aalto.

The new principles of designing the Paimio sanatorium by Alvar Aalto were:

1. Outdoor balconies with good views to the nature give patients a good feeling of well being.
2. From the patient bed you can see nature through big window. To see nature may put your blood pressure down.
3. The connection of patient room window and floor helps cleaning and increase hygienic atmosphere.
4. Wash basin is silent, a new Aalto design helps keep the noise level down.
5. Non–direct ceiling light is not dazzling; you can read books and think.
6. Heating from the ceiling is pleasant, because heat comes all over the space.
7. Natural light can be directed deep to the building corps. Natural light is essential for human beings as for flowers.
8. The acoustic solution to the corridor corner reduce the step noise. It was never built.
9. The yellow colour of the corridor floor is refreshing; colors can stimulate your feelings.
10. Different colors in the ceilings give patients a possibility to admire art.
11. Paimio Chair has wooden ergonomics for TB patients breathing.

These ideas were to heal patients. And so it happened that tuberculosis was conquered in Finland; partly with these design principles, but of course with good antibiotic medication.

这是芬兰首批具有新功能主义特征的建筑之一。因为阿尔瓦·阿尔托出席了欧洲的国际现代建筑协会，国际潮流很快来到了芬兰。而阿尔托通过格罗皮乌斯，特别是柯布西耶，学到了新的建筑理念。

阿尔瓦·阿尔托听取了参与帕米奥结核病疗养院设计的医生们的建议，所以阿尔托也知道心理学的移情作用。

阿尔瓦·阿尔托设计帕米奥结核病疗养院时的新设计原则如下。

1. 建筑外阳台的自然景观视觉效果好，给病人带来舒适的感觉（图8）。
2. 病人可以在病床上通过大窗户看到自然景观，看到大自然可能会让病人的血压下降。
3. 病房窗户与地板的连接设计有助于清洁和空气洁净（图9）。
4. 洗漱盆是静音的，阿尔托的一个新设计有助于降低流水噪声（图10）。
5. 非直射的吸顶灯不会刺眼，病人可以边看书边思考（图11、图12）。
6. 装自建筑天花板的暖气设备令人愉悦，因为热量能遍布整个空间（图13）。
7. 自然光可以直接照到建筑深处。如同对花儿一样，自然光对人类来说也是必不可少的。
8. 走廊拐角处的声学解决方案降低了走在台阶上带来的噪声，但这个方案并没有建成（图14）。
9. 走廊地面使用的黄色让人耳目一新。颜色能刺激人们的感受（图15）。
10. 天花板上的不同色彩为病人提供了欣赏艺术的可能性（图16）。
11. 帕米奥椅的人体工程学设计方便了结核病人的呼吸（图17）。

这些设计想法都是为了治愈病人。就这样，芬兰治好了结核病；一部分是因为这些设计原则，当然主要是因为有好的抗生素药物。

图 8　Outdoor balconies　外阳台

图 9　Patient room window
病房窗户设计

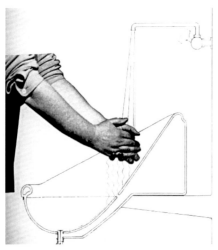

图 10　AA washbasin is silent　阿尔瓦·阿尔托静音
洗漱盆

图 11　Non direct ceiling light
and view　非直射天花板光源和
视觉感受

图 12　Non-direct ceiling light is not dazzling　不产生眩光的散射光源

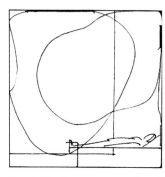

图 13　Heating from the ceiling is pleasant　令人愉悦的建筑天花板暖气设备

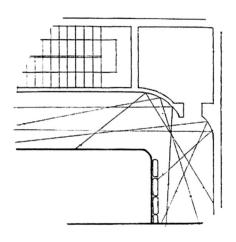

图 14　Acoustic corridor corner　声学降噪走廊转角

图 15　Color at the corridor floor　走廊地面色彩

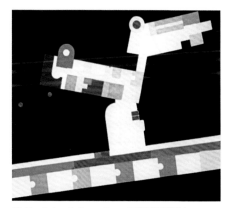

图 16　Colors in the ceiling　天花板色彩设计

图 17　Paimio chair　帕米奥椅

All these details were supposed to heal the circumstances a remarkable way with the patients. Tuberculoses disappeared in Finland by healing. There are no more such sanatoriums in Finland. During the next years this kind of healing details were not used in hospital design in Finland and all the world. It took almost fifty years when new thinking was adopted to hospital design.

Today friendlier, and less machine–like hospital architecture is now promoted as a way to diminish stress. Starting with the way people react to their physical and social environment, the phenomenon of Evidence Based Design that emerged in the United States in the 1980s began to explore how spatial design can influence medical outcomes, with psychology providing the link between personal experiences and objectively verifiable medical data.

Views to nature, access to gardens, patient empowerment (who should be able to control the room temperature, the light , and, obviously have their own flat screen that combines access to television programs and the Internet), clear logistic structures, ample opportunities for social support (leading to the idea to incorporate spaces for family in Intensive Care Units, for instance), noise reduction, provision of daylight (ideally even in operating theatres), a clear preference for single bedrooms, incorporation of elements that give positive distraction—these are only a number of issues propagated on the basis of Evidence Based Design research.

So the healing architecture is coming today to hospital design at last.

Today, the philosophy that guides the concept of the healing environment is rooted in research in the neurosciences, environmental psychology, psychoneuroimmunology, and evolutionary biology. The common thread linking these bodies of research is the physiological effects of stress on the individual and the ability to heal. Psychological environments enable patients and families to cope with and transcend illness.

所有这些细节都能以一种非凡的方式治愈病人。肺结核病在芬兰治愈后消失了，芬兰再也没有这样的疗养院。在之后的岁月里，这些治愈性设计细节并没有在芬兰和世界各地的医院设计中使用。将新的治愈性设计思考纳入医院设计，历经了近50年。

今天，更友好、更非机械化的医院建筑被作为一种减压的方式来宣传。从人们对物质和社会环境的反应开始，20世纪80年代出现在美国的循证设计现象，利用心理学提供的个人经历和客观医学数据之间的关系，开始探索空间设计如何影响医疗结果。

视线可及自然，行走可及花园，赋予病人权利（可以控制房间的温度、光线，以及可播放的电视节目和连接互联网的电视），清晰的逻辑结构，充分的社会支持机会（例如支持重症监护室中一体化的家庭空间），降低噪声，提供日光（理想状态甚至是在手术室中有自然日光），提供单人病房，加入能正面性地分散病人注意力的元素——这些只是基于循证设计研究基础之上引申出的问题。

因此，治愈性建筑最终走向了医院设计。

今天，指导治愈性环境概念的哲学根植于神经学、环境心理学、心理神经免疫学和进化生物学的研究。连接这些学科研究的共同线索是压力对人个体的生理影响和治愈能力。心理环境使患者和家庭能够应对和战胜疾病。

Healing Architecture in Progress

At their time Frank Lloyd Wright and Richard Neutra designed and built houses that had nice views from the main areas of the building.

Also the place for the house was selected like in Feng Shui; not best part of the site, but the best part of the site is for views. The house should open towards south and should be closed towards north. And on a hill, the house should not be built on the top of a hill, but on the slope. And a view towards the sea, lake or river is valuable.

These were also known by Alvar Aalto, when he designed his small houses.

Richard Neutra even wrote a book called *Survival through Design* in 1954, where he is dealing with a possibility of healing architecture.

"A workable understanding of how our psychosomatic organism ticks, information on sensory clues which wind its gorgeous clockwork or switch it this way or that, undoubtedly will someday belong in the designer's mental tool chest."

Physicians and nurses know what happens when a sick person is becoming healthier. The patient's interest is moving from oneself towards the environment, outer world. One pioneer research that was published in Science–magazine in 1984, showed

发展中的治愈性建筑

弗兰克·劳埃德·赖特和理查德·诺依特拉在他们的时代所设计建造的房子，主要区域都有非常好的视野。

房子的选址也是根据风水原理来选择的。不是基址中的最佳位置，而是基址中视线最佳的位置。建筑必须坐北朝南。如果在山上，建筑不应当建在山顶上，而是建在山坡上，朝向为能看见大海、湖泊或者河流。

阿尔瓦·阿尔托在设计其小住宅时，也是遵循这些原则。

1954年，理查德·诺依特拉甚至写了一本名为《通过设计生存》的书，在书中他探讨了治愈性建筑的可能性。

对于我们的身体有机体如何运转，从感官上可以理解为它的复杂精密如华丽发条一般或是类似这样的方式，这无疑是属于设计师的精神工具箱。

医生和护士知道一个病人康复时发生了什么。病人的兴趣点从自己转移到了环境和外部世界。1984年发表在科学杂志上的一项前瞻性研究

that patients (gland cancer surgery) that had a view out of their room to nature were healing faster than the patients that did not have that view. They had a view towards a brick wall.

Roger Ulrich had made that research in 1972–1981. It consisted very exact measurements with quite a big population of patients.

I have also noticed in my practice as an architect, that when you design a room with views to two directions, there is a better atmosphere in that room. And if you have possibility to have views towards to three directions, the better. In the history of modern architecture you can name many examples of this kind of design.

表明，能从房间里看到室外大自然的病人（腺体癌症手术）比无法从房间里看到大自然的病人恢复得更快。后者所看到的只有砖墙。

1972–1981 年期间，罗杰·乌尔里希进行了上述研究。这项研究包括对大量病人群体非常精确的测量数据（图 18）。

作为一个建筑师，在我的建筑设计实践生涯中，我注意到，两个方向都设计有景观的房间，氛围会更好。如果有可能在三个方向都设计有景观，则效果更好。在现代建筑史上，可以举出许多这类设计的例子。

图 18 Espoo Hospital competition entry by K2S Architects 2009
Not realized 芬兰埃斯波医院竞赛入围设计，K2S 建筑事务所
2009，未建成

Environmental Psychology

The basic solutions of the healing architecture come from psychology and environmental psychology. In recent years neuroscience has developed a lot of new information for architects to use in their designs. The knowledge of our brain with magnetic images (fMRI scan) give new information all the time. It is essential to know how our brain is working. Environmental psychology is quite a new area of research. It started in the 1960s in the United States. I studied in North Carolina State University in 1971–1972 where there was courses in environmental psychology. After I came back to Finland, environmental psychology was taken as a curriculum subject in my university for architects in 1973.

During the 1970s architect Kauko Tikkanen did a research for his PhD thesis. In that research, which he started in the USA, he showed that schools that had no windows (only artificial lights) were not so favoured as the schools with windows.

I made my licentiate of technology thesis on *Measurement of Human Responses in Environment* in 1978. I used the old technique of questionnaires to solve the differences with students how they valued street scenes in different situations. I used semantic differential analysis to show the differences. At that time there were no possibility for magnetic images to help this type of research.

治愈性建筑的基本解决方案来自心理学和环境心理学理论。近年来神经学发展了很多新的信息，供建筑师们在设计中使用。利用功能性磁共振成像扫描，发现我们的大脑认识一直在为我们提供新的信息。了解我们的大脑如何工作是非常重要的。环境心理学是一个比较新的研究领域，始自 20 世纪 60 年代的美国。1971–1972 年期间，我在美国北卡罗来纳州立大学学习，那里开设有环境心理学课程。1973 年，我回到芬兰后，我所在大学的建筑系开设了环境心理学课程。

20 世纪 70 年代建筑师卡奥古·蒂卡宁做了一项博士论文研究。在这项从美国开始的研究中，他发现没有窗户（只有人工照明）的学校不如有窗户的学校受欢迎。

1978 年，我完成了执业资格关于《测量人体在环境中的反应》的技术论文。我用问卷调查的老方法来解决学生们在不同情况下评价街景的差异。我使用了语义差异分析来列出这些差异，那时还没有磁共振成像扫描图像来协助这一类型的研究。

During the 1980s, when I left the department, teaching environmental psychology for architects was canceled. That was a pity, because psychology is a very important part of architectural education.

It is still important to know the relationship with a human being and the environment, because with that knowledge we could design better buildings for human beings. And we could get more information how people use their buildings and environment. At the same time we could depend our knowledge of empathy.

Empathy is one of the key words for architects who design for human beings.

20 世纪 80 年代，在我离开建筑系时，环境心理学课程被取消了。这很令人遗憾，因为心理学是建筑学教育中一个非常重要的组成部分。

了解人与环境的关系仍然很重要。有了这些知识，我们才能为人类设计更好的建筑，能够得到更多人如何使用他们的建筑和环境的信息。同时，我们也可以加深对同理心的了解。

同理心是为人类做设计的建筑师的关键词之一。

Neuroscience

Already PrimeMinister Winston Churchill in UK made it clear over half a century ago that "We shape our buildings and thereafter they shape us".

Now it is evident by neuroscience that our environment has an effect on our brain. A specialized neuron called a mirror cell was discovered in 1991 by four Italian scientists in Parma University. These special neurons can subconsciously imitate things that are done by other human beings or animals or how they feel. This discovery can give new sight to our behavior and mental understanding of emotions, state of mind and empathy.

The first conference together with Architects and Neuroscientists were held in Woods Hole near Cape Cod in USA 2002. In that conference Roger Ulrich's research on hospital views in 1984 was discussed thoroughly.

Italian physiologist Angelo Mosso was a pioneer in brain research. In 1882, he built the first neuroimaging device: precision scales on which the test subject rested. Mosso proved that when a person thinks, their head drops lower. Reading a philosophical book had a more substantial effect than skimming a newspaper. From this, Mosso deduced that the brain requires more blood when it works harder.

早在半个多世纪前，英国首相温斯顿·丘吉尔就曾明确表示："我们塑造了自己的建筑，随后建筑也塑造了我们。"

当今神经科学证实，我们的环境会影响我们的大脑。1991 年，帕尔马大学的四位意大利科学家发现了一种叫作镜像细胞的特殊神经元。这些特殊的神经元能够下意识地模仿其他人或动物所做的事情或是感受。这一发现可以为理解我们的行为和情绪、心态、移情等精神状态提供新的视角（图 19～图 22）。

2002 年，在美国科德角附近的伍兹霍尔举行了建筑师和神经学家共同参加的第一次会议。这次会议深入讨论了罗杰·乌尔里希 1984 年发表的关于医院研究的观点。

意大利生理学家安杰洛·莫索是大脑研究的先驱。1882 年，他发明了第一台神经成像设备：精确测量受试者的量表。莫索证实，当一个人思考时，他的头会垂得更低。阅读一本哲学著作比浏览一份报纸对人的影响更持久。由此，莫索推断出大脑工作越努力，越需要血液。

Brain activity is the movement of molecules and atoms, chemical reactions, electricity and magnetism; as measuring technology has developed, we have gained more delicate and detailed ways of peeking inside our heads: CT, DBS, DOI, DOT, DTI, ECOG, EEG, EROS, FMRI, HD.DOT, MEG, MRI, NIRS, PET, SPECT, TES, TMS...

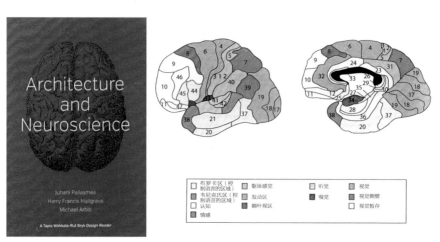

图19 A view of left cerebral cortex showing Brodmann's numbered areas with some color coding of functions Broadmann 大脑分区，根据功能进行颜色编码的左侧大脑皮质图像

大脑活动是分子和原子的运动，也是化学反应与电磁反应。随着测量技术的发展，我们获得了更精确和更详细的方法，以此来窥探我们的大脑：如 CT、DBS、DOI、DOT、DTI、ECOG、EEG、EROS、FMRI、HD.DOT、MEG、MRI、NIRS、PET、SPECT、TES、TMS 等。

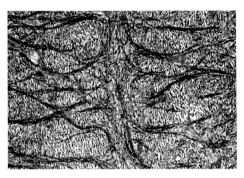

图 20 Nerveweb 神经网

图 21 Brain 大脑

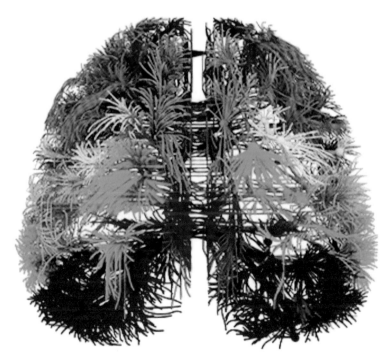

图 22 Connectome 连接体

Some techniques, such as PET imaging, are commonly used when examining the head and body in hospitals, whereas other methods are still being tested. The earliest method, EEG measuring brain activity, became a mail order hit when hippies striving to find enlightenment started to measure their theta waves with EEG equipment.

As incredible as it sounds, we can even watch thoughts and dreams—those of other people. Ultimately, the principle is fairly simple. The measuring device registers your brain activity when you focus and stare at a pair of scissors on a table. When you move your eyes to a coffee mug, your brain reacts a little differently. Are you now thinking of the scissors or the mug? The machine reads your thoughts.

Jack Gallant, a professor at University of California Berkeley, had young people in his research group lie still in an fMRI scanner and watch countless snippets of videos collected from the internet. In fact, the scanner does the same as Mosso's scales: measures blood flow in the brain—by observing the oxygen molecules in blood. fMRI images indicate the parts of the brain where something happens: here, here and there when the student watches a video of a bird flying. A computer analyses tens of thousands of images and compares them to the videos, learning slowly to connect the fMRI image and the distribution of colourful areas to different photos. In the end, only the result of the fMRI imaging is entered to the computer and it tries to reconstruct what the subject saw.

In the next ten years, Americans want to chart each and every neuron of the human brain whereas Europeans want to learn how to simulate brain activity using a supercomputer. The final destination is the same: to understand the activity and structure of the brain so well that we can genuinely start to do something about our mental disorders, maladies and degeneration.

Professor Roger Ulrich, who is one of the most prominent researchers in the field of understanding the psychological and physiological impact of physical environments on the healing process undertook a comprehensive review of some 600 peer–reviewed research studies on the design of health buildings. The studies covered aspects of hospital design such as the impact of improved daylighting

有些技术，例如 PET 成像，通常被使用在医院中做人体头部和身体检查时，还有一些检测方法则还在测试阶段。早期，由于利用脑电图测试大脑活动，嬉皮士们开始用此来测量他们的西塔（θ）波，这一设备成了他们邮购的热门。

这听起来难以置信，我们甚至可以观察其他人的思想和梦境。归根结底，其原理相当简单。当你专注地盯着桌上的一把剪刀时，这种测量设备会记录下你的大脑活动状况；当你将注意力转移到一个咖啡杯时，你的大脑会做出不同的反应。你现在想的是剪刀还是杯子，可以从设备中读取。

加州大学伯克利分校的杰克·加兰特教授，让他研究小组中的年轻人静静地躺在功能性磁共振成像扫描仪中，观看无数从互联网上收集来的视频片段。事实上，这台扫描仪的功能与莫索的神经成像设备一样：通过观察血液中的氧分子来测量大脑中的血流量。功能性磁共振成像扫描仪图像显示出，当学生观看一只鸟飞行的录像时大脑中发生变化的部位。计算机能分析成千上万的图像，并将它们与视频进行比较，慢慢学会了将功能性磁共振成像图像和大脑中分布的不同彩色区域与不同图像联系。最后，只要有功能性磁共振成像的结果输入计算机，计算机就能试着重建扫描对象所看到的对象。

在接下来的 10 年，美国人想要绘制人脑的每一根神经元，而欧洲人想要学习如何用超级计算机模拟大脑活动。最终目的一样：为更好地了解大脑的活动和结构，以便我们能真正地开始针对人们的精神障碍、精神疾病和神经退行性疾病做一些事情。

罗杰·乌尔里希教授是物理环境对康复过程的心理和生理影响领域最著名的研究者之一，他对大约 600 项经同行审查的关于医疗建筑设计的研究进行了全面详细的研读。这些研究涵盖了医院设计的各个方面，比如改善采光水平和病人视野，控制噪声，使用单间病房而不是多床

levels and views for patients, the impact of improved noise control, the use of single–rooms as opposed multibed wards, design to reduce cross–infection, and improved layouts for staff, patients and visitors. The conclusions were published by Ulrich and others in 2004 in a document titled *The Role of the Physical Environment in the Hospital of the 21ˢᵗ Century.* Based on that research, it seems that the single–rooms are coming to hospitals in the future.

We in Finland have had some seminars on neuroscience at Tapio Wirkkala and Ruth Bryk–Foundation with published seminar books in recent years. These books are published by the initiative of architect and professor Juhani Pallasmaa, who is the most famous phenomenologists in Finland today.

Phenomenology is a philosophical trend of environmental perception that has been based on the ideas of one of Hegel, Husserl, Heidegger, Merleau–Ponty, Derrida and others.

Although generally preferring the term poetic to aesthetic, phenomenology is the dominant mode in architecture where the question of sense and affect is posed. It proposes to explain directly how the spaces we inhabit make us feel. The idea that architecture can be made better through understanding our perception of space and putative spatial archetypes began in the eighteen century, took its modern form in the late nineteenth century in empathy theory, grows in empirical psychology in the mid twentieth century and is reviving today in an uptake of neuropsychology.

The ideas and historical trajectory of architectural phenomenology are best understood as a battle for intellectual hegemony in the academy. Paradoxically, this sensibility for direct experience requires cultivation through a highly theorized account of architectural history. Phenomenology nevertheless remains the strongest case we have of architectural thinking claiming the authority of a fundamental outside, something beyond the particular concepts of art or architecture specific to a culture of time.

病房，减少交叉感染的设计，以及改进工作人员、病人和探视者的布局设计等所产生的影响。2004 年，这些结论由乌尔里希等人在一篇名为《21 世纪医院物理环境的作用》的文献中公开发表。基于这篇研究，看起来单人病房会在未来逐步进入医院。

在芬兰，我们在塔皮奥·维卡拉 – 露丝·布吕克基金支持下举办了一些关于神经科学的研讨会，最近几年还出版了研讨会的书籍。这些书籍是由当今芬兰最著名的现象学家、建筑师尤哈尼·帕拉斯玛教授提议出版的。

现象学是一种以黑格尔、胡塞尔、海德格尔、梅洛 – 庞蒂、德里达等人的思想为基础的环境感知的哲学化趋势。

虽然现象学更倾向诗学而非美学，但现象学是建筑中提出感觉和情绪方面问题的主导模式。建筑现象学试图直接诠释居住空间如何给人们以不同感觉。通过理解我们对空间的感知和假定的空间原型，建筑可以被建造得更好，这一想法始于 18 世纪，19 世纪晚期的移情理论使之发展成为现代形式，伴随 20 世纪中期的经验心理学逐步发展，如今在神经心理学的发展中复兴。

建筑现象学的思想和发展历程最容易被理解为一场学术霸权之争。矛盾的是，这种直接体验的敏感性需要通过高度理论化的建筑历史来培养。然而现象学依然是我们所拥有的最强有力的关于建筑学思想的案例，它宣称一种基本的外在权威，超越了特定于一种时代变化的艺术或建筑概念。

Christopher Alexander

I want it to be possible for us to make
buildings, which have that simple
comfort in them, so that everyone
feels at home, so that they support us
in our daily life.

—Christopher Alexander

American architect Christopher Alexander has tried to solve the problem of our
tradition to build that did not continue anymore from the 1960s. His first famous
research was named "A City is not a Tree", where he proposed a compact city
for people; a kind of matrix. Cities were built mostly with the tree–like road
systems around the world. *A Pattern Language* (1977) is the most known of his
books. In that book he has made a list of design patterns from the tradition and
some ones too to make a better environment. He has tried to give advice how to
place your house on a lot, how you design the plan and even small details. Most
of these design solutions were used in traditional buildings, but knowledge has
disappeared during time.

The newest book by Alexander is called *The Nature of Order* (2003). It has four
books: *The Phenomenon of Life, The Process of Creating Life, A Vision of a
Living World* and *The Luminous Ground.*

This book is a synthesis of his earlier books and ideas, like *The Search for a
New Paradigm in Architecture* (1983) and *The Timeless Way of Building* (1979).
The main part of the book consists ideas how he thinks that life is in the nature
in different ways. Some environments are designed with life, some are designed
dead or disgusting. His ideas are sharp, but he cannot say that to make good

克里斯多夫·亚历山大

我希望我们能建造出舒适的建筑，让每个人都有宾至如归的感觉，以便让人们在日常生活中支持我们。

——克里斯多夫·亚历山大

美国建筑师克里斯多夫·亚历山大试图解决传统建筑在 20 世纪 60 年代后出现的断层问题。他第一个著名的研究名为"一座城并非一棵树"（图 23），在这一研究里，他给人们描绘了一座布局紧凑的城市，类似一个矩阵的城市。那时，世界各地的城市多是以树状的道路体系构架建造的。1977 年出版的《建筑模式语言》是他最著名的著作。在这本书里，他列出了一系列来自传统的设计模式，以及一些用来创造更好环境的设计模式。他试着在房子如何选址、如何设计，包括设计细节等方面给出建议。这些设计方案曾用于传统建筑，但随着时间的推移，这些建筑设计知识已经逐渐消失。

亚历山大最新的著作是"秩序的本质"系例（2003）。其中包括四本书：《生命的现象》《创造生命的过程》《生机世界图谱》和《发光的地面》。

这套著作是他早期著作和思想的集合，如《建筑新范式的探索》（1983）和《永恒的建筑之道》（1979）。这套书的主要内容是生命如何以不同方式存在于大自然中（图 24～图 27）。有些环境设计是与生命结合的，有些则是毫无生机或令人厌恶的。他的观点非常敏锐，但他不能说为

architecture you need a human architect to do it. Architecture needs empathy. The architectural works by Alexander himself shows, how big the cap between good design and good research may be. We need more research by architects too, to be able to design healing architecture.

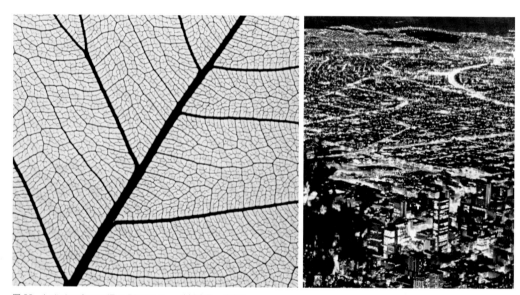

图 23　A city is a forest (Esa Piironen)　一座城市是一片森林

图 24　Wave forms　波浪形式　　图 25　Snow　雪花

了做出好的建筑，需要一个人类建筑师。建筑需要同理心。亚历山大自己的建筑作品，让我们看到好的设计和好的研究之间的差距有多大（图 28）。我们也需要建筑师有更多的研究，才能设计出治愈性建筑。

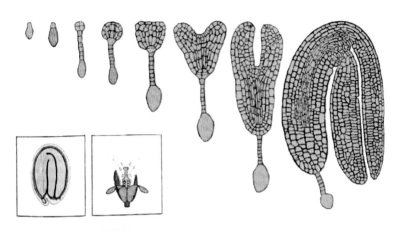

图 26　Growing plant　植物生长

图 27　Easy going　友好型环境

图 28　Architecture of Alexander　亚历山大的建筑作品

Anthroposophy

Anthroposophic medicine was developed out of indications of Rudolph Steiner (1861–1925) at the beginning 20th century. Steiner built his Goetheanum center to develop anthroposophic methods in Germany and his schooling system is famous worldwide. Steiner's ideas were obtained by a Danish–born architect Erik Asmussen (1913–1998), who started to build a community in Järna Sweden from 1985 to heal people.

Anthroposophic medicine includes allopathic medicine, it makes use primarily homeopathic and rhythmic massage, mineral baths, proper nutrition and a range of artistic therapies such as painting, song and eurythmy. Erik Asmussen has built there lot of buildings, where he has used his ideas of healing architecture. Colors are a vital part of his architecture. Color is never used for merely decorative purposes; rather it is intended to serve the process of healing and support the functions of spaces in which it is applied.

Occasionally, exterior walls inflect outward, creating asymmetrical window bases and alcoves. Windows are designed with great care to allow views from each bed location and far enough so that the patient lying in bed can actually look outdoors and see more than the sky.

Asmussen's architecture is based on seven principles of healing architecture.

人智学

人智医学是在 20 世纪初，以鲁道夫·斯坦纳（1861-1925）创建的人智学为基础发展起来的。斯坦纳在德国建立了歌德堂来发展精神分析的方法，他创建的教育制度世界闻名。斯坦纳的想法来源于丹麦建筑师埃里克·阿斯穆森（1913-1998），自 1985 年起，阿斯穆森在瑞典的雅纳建立了一个社区来治疗人们（图 29）。

人智医学包括对抗疗法，主要使用顺势疗法和节奏按摩法、矿泉浴、适当营养疗法，以及一系列的艺术疗法，如绘画、唱歌和音语舞等。埃里克·阿斯穆森建造了很多建筑，在这些建筑中，他运用了其治愈性建筑理念。色彩是其建筑的重要组成部分。颜色从来不只是用来装饰的，相反，其目的是服务于治愈过程，以及支持运用这一色彩的空间的功能。

偶尔，外墙向外弯曲，形成不对称的窗基和凹槽。窗户精心设计，以使每张病床的视线足够远，保证病人躺在床上可以看到户外的远处，而不仅仅是天空。

阿斯穆森的建筑基于以下七条治愈性建筑设计原则。

1. Unity of form and function is sometimes called spiritual functionalism.

2. Polarity implies that differences are not merely oppositions, but the differences are part of a larger whole. Asmussen is trying to design from the principles that generate form in nature rather than the products that nature produces.

3. Metamorphosis concept is the most important and the most difficult to grasp. It is based on Johannes Goethe's studies of the flowering plant, when nature continually transforms itself.

4. Harmony with nature and site. Asmussen's buildings are visibly shaped by and in conversation with naturally occurring features nearby.

5. The living walls are organic and express the generative forces by which they are shaped.

6. The dynamic equilibrium of spatial experience. Asmussen has two primary kinds of spaces—those for movement and those for rest. In the movement spaces, there are places for rest, and in the spaces for rest, which are typically specific activities like sculpture or painting, there is also a sense of movement. He uses contrasts between in and out, front and back, near and far to create spaces that are dynamically polarized and alive.

7. Color luminosity and color perspective. The color on and in Asmussen's buildings enlivens and reveal materials, which always remain visible beneath its glow. He uses a transparent method in coloring his buildings.

The architecture of Asmussen is considered as original, maybe organic, but universally it is not among the master pieces of modern architecture. But in a way it is one good example of architecture that can show us a new way towards healing architecture.

1. 形式与功能的统一有时被称为精神功能主义。

2. 两极性意味着差异不仅仅是对立，而且是一个大整体的一部分。阿斯穆森的设计试图遵循从自然中产生形式的原则，而不是自然产生的产品。

3. 变形的概念最重要，也最难以把握。它基于约翰尼斯·歌德对大自然不断改变自身情况下开花植物的研究。

4. 与自然和场所和谐共处。阿斯穆森的建筑明显地由附近的自然特征塑造，并与之形成对话。

5. 绿植围墙是有机的，表现着塑造形态的生命力。

6. 空间体验的动态平衡。阿斯穆森的建筑有两个基本的空间类型——运动空间和休息空间。在运动空间中有休息的地方，而在休息空间中，也会有特定的活动处，如雕塑或者绘画，有一种动感。他利用内外、前后、远近之间的对比，创造出动态、两极分化、充满活力的空间。

7. 颜色亮度和颜色透视。阿斯穆森建筑内外的色彩使材料显得生动活泼，在颜色的光度下始终保持醒目。他用一种透明的方法为建筑物着色。

阿斯穆森的建筑被认为是原始的，或许也是有机的，但通常来说，并不属于现代建筑大师作品。然而，在某种程度上，他的建筑是向我们展示治愈性建筑一种新方式的极好案例（图29）。

图29　Järna　瑞典雅纳社区建筑

Maggie's

An American architect and critic Charles Jencks (1939–2019) who lived in London has made his contribution to help building healing architecture in the memory of his wife Maggie that was passed away with cancer some years ago.

These small rehabilition centers for the patients of cancer have been built by the money that has been collected privately around British Isles. Jencks has asked some of the world best architects to design these houses: Frank Gehry (Dundee 2003), Zaha Hadid (Fife 2006), Rogers Stirk Harbour + Partners (London 2008), Kisho Kurokawa (South West Wales 2011), Rem Koolhaas (Glasgow 2011). Buildings have been designed as homelike as possible with furniture from the flee market. There is no research made yet to tell how well these buildings heal: These houses do not heal cancer of course, but they help the patients to participate in conversations and meetings to make the life meaningful once more. New buildings are going to be built in Oxford, Hong Kong and Barcelona.

In that little book *Can Architecture Affect Your Health* Jencks reminds us architects about a research made in the 1920s in a factory called Hawthorne in Chicago. Researchers wanted to solve, how the better lightning in factory halls would increase the amount of productivity. When they added more lights in the factory halls, the productivity was increased. Later the researchers found that it was not the lightning that was causing the more productivity, but the thing that it

麦琪医疗中心

生活在伦敦的美国建筑师、建筑评论家查尔斯·詹克斯（1939–2019），为了纪念他数年前因癌症去世的妻子麦琪，在治愈性建筑方面做出了他的贡献（图30～图32）。

这些为癌症病人设立的小型康复中心是用不列颠群岛各地私人募集的资金建立起来的。詹克斯邀请世界上最优秀的建筑师来设计这些建筑：弗兰克·盖里（邓迪，2003），扎哈·哈迪德（法伊夫，2006），罗杰斯·施蒂克·哈珀与合伙人事务所（伦敦，2008），黑川纪章（西南威尔士，2011），雷姆·库哈斯（格拉斯哥，2011）。这些建筑师设计的建筑做到了尽可能像家一样，家具都是从跳蚤市场淘来的。目前还没有研究来说明这些建筑在治愈性上的优势，在英国牛津、中国香港和西班牙巴塞罗那都将兴建这样的康复型建筑。它们当然无法治愈癌症，但却有助于病人参与对话与会议，使生活再次变得有意义。

詹克斯在《建筑会影响你的健康吗》一书里提到20世纪20年代在芝加哥一家叫霍索恩的工厂里做的一项研究。研究人员想要解决的是怎样改善照明，以提高产品产量。当在厂房增加更多灯时，产量便提高了。后来研究人员发现，并不是灯光导致增产，而是当工人第一次被问及

was the first time that the workers were asked anything about their circumstances in the factory. They felt that the leaders in the factory cared for them. This phenomenon is called the Hawthorne–effect.

It is the same effect in architecture when the inhabitants are taken as members of the design team when designing housing. Some architects have done this with good results.One of them was a swedish–british architect Ralph Erskine (1914–2005).

图 30　Maggie's Nottingham, Pier Gough, 2011　诺丁汉麦琪医疗中心，皮耶尔·高夫，2011

图 31　Maggie's Highlands, David Page, 2005　高地麦琪医疗中心，戴维·佩奇，2005

他们在工厂所处的环境是否有助于提高产品产量时，工人感受到了来自领导的关心。这个现象被称作霍索恩效应。

在设计住宅时，将居住者作为设计团队的一员，在建筑学上也有同样的效应。一些建筑师已这样做，并取得了良好的成效。其中一位是瑞典裔英国建筑师拉尔夫·厄斯金（1914–2005）。

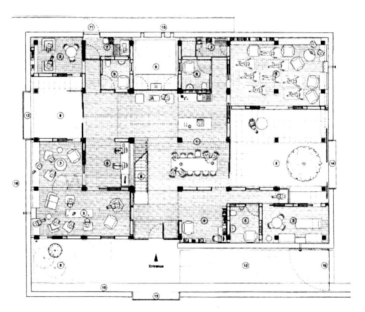

图 32　Maggie's London, Rogers, Stirk, Harbour + Partners, 2008
伦敦麦琪医疗中心，罗格斯建筑事务所，2008

Elements of Healing Architecture

Architecture is experienced by human beings with five senses: visually, hearing, touching, tasting and smelling. Somebody has said that we even have more senses, for instance equilibrium, kinesthetic sense and so on. But it's essential that we architects design for these five senses as well as possible and with empathy.

The universally recognized great places, spaces and buildings created throughout the history of civilization are not admired, visited, envied and enjoyed by millions for the quality of the project management, value engineering or scheduling used in their creation, but the applied creativity in the manipulation of scale, mass, form, volume, solid and void, light and shadow, material, texture, color, harmony, art and landscape. Through the application of such skills, places are created that we intuitively know are right, that lift our spirits, bring delight to our souls and imbue us with feelings of well–being. That is what great architecture created by great architects can do.

The impact of the quality of our environment on our health and well–being is hugely significant, be it in relation to larger urban spaces, housing, streetscapes, work places, individual houses, right down to the quality of the individual spaces and rooms we occupy within buildings.

治愈性建筑的设计元素

人们利用五感即视觉、听觉、触觉、味觉和嗅觉去体验建筑。有些人认为人们还有更多的感受，如平衡感、动感等。但我们建筑师尽可能以同理心考虑五感设计，这是最重要的。

在整个文明史上举世公认的伟大场所、空间和建筑，受到千百万人的仰慕、参观与喜爱，并成为人们的享受之地，这不是因为建造过程中高质量的工程管理、有价值的工程技术或者合理的时间进度，而是在尺度、质量、形式、体量、实与虚、光与影、材质、纹理、色彩、和谐、艺术和景观中，创造力发挥合力的结果。通过这些技能的应用，我们可以创造出通过直觉感受认为适合的场所。这些场所可以振奋我们的精神，给灵魂以愉悦，并使我们充满幸福感。这就是伟大建筑师创造出的伟大建筑所能做到的。

我们的环境质量对我们健康和幸福的影响至关重要。它关系到更大的城市空间、居住建筑、街道景观、工作场所、个人住宅，甚至包括我们在建筑中所占有个人空间的质量。

Medical research has clearly demonstrated that the immune systems of all people are already under pressure and further stress, including that brought about as a result of the environments in which they are being cared for, further weakens their ability to resist illness or fight infection.

Equally research demonstrates that good design in the environment can positively impact on our immune systems and powers of recovery as well as helping to eliminate potential risks to the well–being of patients for instance in hospitals.

That's why I try to formulate some basic factors that should be considered when trying to put forward new parameters towards healing architecture.

Home is usually the best place for human beings. It is like a nest for birds and other animals. To design homelike places is one goal for architects.

I have mentioned here just a few elements for healing architecture. Here are a lot more details for healing architecture that we have to take care of.

To generate more goals and means for architects to design healing architecture is a continuous and movable task. It is not an easy goal to achieve. The new paradigm in architecture is maybe difficult to realise, but I think it is better to put our goals high and not to reach them than to put them too low and reach them.

Vitruvius' three tenets of architecture – firmitas (solid), utilitas (practical) and venustas (beautiful) – would get one more quality: curare (healing), in latin.

Light

Eighty percent of what we interpret of our surroundings comes to us from what we see of our environment and that is greatly affected by the light available in that environment. Lighting design in healthcare environments is a major factor in creating healing situations. Since the design of healthcare environments is

医学研究清楚地表明，所有人的免疫系统都已处于一定的压力下，并面临更大的压力，包括他们所处的护理环境所带来的压力，这些都进一步削弱了他们抵抗疾病或感染的能力。

同样，研究表明，例如在医院，良好的环境设计可以对我们的免疫系统和康复能力产生积极影响，并有助于消除影响患者健康的潜在风险。

这就是为何我在试图提出治愈性建筑的新参数时，先试着制定一些基本因素。

家通常是人们最爱的场所。如同鸟巢之于鸟，或巢穴之于动物。设计家一样的建筑，是建筑师的设计目标之一。

我在这里只提到了一些治愈性建筑的元素，但还有很多治愈性建筑的细节等着我们去处理。

以更多的设计目标和手法设计治愈性建筑成为设计师持续与不断变化的任务，这不是一个容易实现的目标。这种建筑的新范例也许很难实现，但我认为设定较高且不易实现的目标，要比设定可轻易实现的过低目标好得多。

维特鲁威的建筑三原则——坚固、实用、美观——需要再增加一个原则：治愈。

光

我们对周边环境的解读 80% 以上来自我们所看到的环境，而这在很大程度上受到环境中光线的影响。医疗环境中的照明设计是创造康复情境的主要因素。有说法说医疗环境设计会影响患者的康复情况，但高昂的花费使医院不愿意改造或者重修，因此改变照明成为改善现有环

said to influence patient's outcomes, yet high costs prevent most hospitals from renovating or rebuilding.

Changes in lighting becomes a cost-effective way to improve existing environments. It is proven that people who are surrounded by natural light are more productive and live healthier lives. When patients are sick, and surrounded by medical equipment and white walls, the last thing they need is dark, stuffy room. This is why it is important for every room to have a window for natural light to come into and help create a healing environment for the patient.

Daylight is a crucial source of energy in our life. It has significant effects on human beings both psychological and physiological. The effect of light on our circadian rhythm (i.e. biological systems that repeat 24 hours) has been recognized for many years. Some suggests that light is the most important environmental input in controlling bodily function after food (La Grace 2014). Several researchers agree that lighting has a profound effect on human's hormonal and metabolic balance.

Most researchers believe that daylight has a significant effect on work productivity in an office environment. A study of worker productivity level found that daylight may positively affect the work performance in an interior or windowed office environment during winter months. However the quantitative relationship of daylight and productivity has not been established. Another study indicates scientifically that lighting systems (i.e. a combination of daylight and artificial light) appear to be important for both visual performance and biological simulation i.e. circadian rhythm. It also concludes that human alertness, work performance, sleep quality and degree of comfort and well being are adversely affected by inadequate light.

境的一种经济有效的方法。

事实证明，被自然光环绕的人们更有活力，生活更为健康。当人们生病，被医疗设备和白墙所包围时，他们最不想要的就是黑暗、沉闷的房间。这就是为什么每个房间都有窗，让自然光进入窗户是如此重要，能为病人创造一个治愈性的环境。

日光是我们生活中重要的能量来源。它对于人类的生理和心理都有重要的影响。我们认识到光对人的昼夜节律（即 24 小时重复的生物系统）的影响已有多年。有些人认为，光是继食物之后控制身体机能最重要的环境输入（拉·格雷斯，2014）。数项研究认为，照明对人体的激素和代谢平衡有重要的影响。

绝大多数研究者相信，日光对于办公环境的工作效率有非常重要的影响。一项关于工人工作效率水平的研究发现，在冬季的几个月里，在室内或有窗的办公室环境中，日光可能会对工作表现产生积极的影响。然而，光照和生产效率之间的定量关系尚未建立。另一项研究表明，光照系统（即日光和自然光的结合）似乎对视觉表现和生物钟（即昼夜节律）都很重要。这项研究还得出结论：光线不足对人的警惕性、工作表现、睡眠质量和舒适程度及身体状态都有不利影响（图 33）。

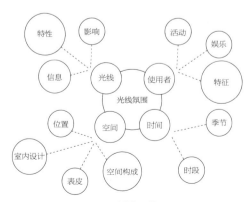

图 33　The Model of Light Atmosphere　光线氛围模型

Conversely, psychiatrists generally agree that the absence of daylight may cause several health disorders. These are sadness, fatigue, mood disorders and seasonal affective disorders.

Ulrich's research provides evidence supporting something we all intuitively know: the natural therapeutic, recuperative and restorative effects of daylight. All of us have experienced that sensation of energy flowing into us and our spirits being lifted when we feel the touch of warm sunlight on our faces.

Despite this, when dealing with the group of people most in need of the tonic that is natural daylight, amazingly many health buildings are still designed with deep plan solutions, internal rooms, enclosed waiting spaces and what I refer to as the tyranny of double loaded permanently artificially lit corridors, thus depriving occupants of this miraculous free element of natural daylight.

Danish Lone Stidsen has made a research on *Design Parameters of Pleasurable Light Atmospheres in Healing Hospitals*. Light affects people in many ways and light can be used to create a certain atmosphere in a particular situation. Based on Gernot Böhmes'concept of atmosphere she has produced a Model of Light Atmosphere to be used in design. It means, that we should create a "home–like" and "pleasant or appealing" light atmosphere for our hospitals. The light should be chosen on the basis of the patients'needs for that feeling. This will in most cases mean a color temperature that is higher than 3000K and a reasonably good color rendering of Ra over 80. Reflecting on the importance of light as the fundamental of architecture Louis Kahn wrote:

"The sun never knew how great it was until it hit the side of a building". Kahn himself an expert manipulator of form and light, was reminding us that it is light that gives life to architecture and our experience of it. He clearly recognized the

神经科医生普遍认为，日照的缺失可能导致一系列健康障碍，包括悲伤、疲劳、情绪障碍和季节性情感障碍等。

乌尔里希的研究为我们凭直觉知道的事情提供了支持证据：日光具有自然治疗、恢复和复原作用。我们所有人都体验过当温暖的阳光照耀在我们脸上时，能量流进我们的身体，精神随之振奋的感受（图34）。

图 34　Main atrium of Pikku-Huopalahti Multipurpose House, Helsinki, Architect:Esa Piironen,1997　皮库 - 霍帕拉赫蒂社区中心主中庭，赫尔辛基，建筑师：埃萨·皮罗宁，1997

尽管如此，令人惊讶的是，当面对一群最需要自然光的人，许多医疗建筑仍旧采用大进深建筑方案，内向的房间、封闭的等候空间和如同暴政一般增加双重负担的永久人工照明走廊，从而剥夺了空间使用者免费享受自然光这种元素的权利。

丹麦人隆·斯蒂德森做过一个关于《医院舒适光环境的设计参数》的研究。光线在许多方面影响着人们，在特殊环境下还可创造出特别的氛围。基于格诺特·博梅斯（Gernot Böhmes）的氛围概念，她制作了一个用于设计的光线氛围模型。这意味着，我们应该为医院创造一个"家一般""令人愉悦而有吸引力"的光线氛围。光线的选择应当基于患者的感受所需。大多数情况下意味着光线色温高于3000K和高于80的相当好的显色指数。

当谈到将光作为建筑之本思考光的重要性时，路易斯·康（Louis Kahn）写道："太阳从不知道它有多伟大，直到它遇到建筑立面。"康是一位运用形式与光线的高手，他一直提醒我们，光赋予了

importance of daylight, not jus for its impact on the exterior of buildings, but also on those inside the building. He also said, "A room is not a room without natural light".

There is not enough research made to investigate the meaning of natural light for human beings. People living in the northern hemisphere get very little natural light during the wintertime. That's why we should build more glass roof spaces to get more light daily. We as human beings are like plants that need natural light and sunlight. The excess of light is easy to control, but the lack of natural light causes even illnesses. Light is the most noble material in architecture.

Noise

Noise is harmful to people. Research has shown that people who are continuously exposed to noise develop different kinds of diseases. Noise increases stress that is harmful to people. The never–ending muzak of shopping centres is a nightmare.

Sound is defined as a longitudinal wave motion that advances in liquid, gas or solid matter. It contains wavelengths that are perceptible by humans, and it is sufficiently loud to be heard. Qualities of sound include volume, frequency and spectrum. Spectrum consists of the division of sound energy along different frequencies. Volume, or sound pressure, is expressed using a logarithmic decibel scale. People's susceptibility to different frequencies varies. That is why A–frequency weighting is used when estimating a sound level as it expresses the volume of the sound we hear. The following table includes examples of different volumes (the decibel levels in the table are approximate):

–dB 0 auditory threshold
–dB 10 gentle rustling of leaves
–dB 20 watch ticking
–dB 30 whisper

建筑生命力以及我们对建筑的体验。他清楚地认识到光线的重要性：不仅是因为它对建筑物外部有影响，还因为它对建筑内部空间也有影响。他还说："没有自然光线的房间称不上房间"。

还没有足够的研究来调查自然光对于人类的意义。生活在北半球的人们在冬天能享受到的自然光非常少。这就是为什么我们要建造更多的玻璃屋顶空间，以便每天得到更多的光照（图 34）。人类和植物一样需要自然光。光线过多容易控制，而光线不足甚至能导致疾病。光是建筑中最高贵的材料。

噪声

噪声对人类有害。已有研究表明，经常暴露在噪声中的人会患上各种疾病。噪声会增加对人体有害的压力。购物中心永不停息的音乐简直就是噩梦。

声音是一种在液体、气体或固体中移动的纵波。它包含了人类可以感知的波长，并且足够响亮，可以被人类听到。声音的特性包括音量、频率和频谱。频谱由声能沿不同声音频率的分布组成。音量或声压用对数分贝来表示。人们对不同频率声音的敏感度是不同的。这就是为什么在估计一个声音水平时，使用 A 频率加权来表达我们听到的声音音量的原因。下表中包含了不同分贝音量的例子（表中的分贝水平描述为近似值）。

0 分贝	听觉阈限
10 分贝	树叶轻轻的沙沙声
20 分贝	手表的滴答声
30 分贝	悄悄话

–dB 40 bird song	
–dB 50 office noise	
–dB 60 discussion	
–dB 70 traffic noise on the street	
–dB 80 vacuum cleaner	
–dB 90 orchestra playing	
–dB 100 chainsaw	
–dB 110 air drill	
–dB 120 jet airplane	
–dB 130 pain threshold	

On average, sounds exceeding 100 decibels are experienced to be uncomfortably loud, and a sound pressure level of 110–130 dB causes a sensation of pain. Sounds of nature can span from a barely perceptible rustle of a leaf to a roar of thunder exceeding 100 decibels.

As sound vibrates, it creates the sensation of hearing. The inner ear senses the sound waves, and the actual sensation of the sound is created in the auditory cortex. The auditory pathways are connected to the central nervous system and especially to the limbic system that control our emotional actions. This means that auditory perceptions are processed through emotions, memory and attention. All sounds we are exposed to do not make their way to our consciousness. This is because the central nervous system processes the information entering through the ear. If we classify a sound as having little significance, we don't pay attention to it. On the other hand, a new, previously unknown sound or a sound we identify as important based on our previous experiences makes us alert. In these cases, our auditory pathways focus on listening to this signal. We become especially alert with sounds that we experience to be new and threatening. The auditory pathways aim to strengthen the perception of such sounds. Sounds that have been identified as important may cause anxiety, fear or other negative emotions.

40 分贝	鸟儿鸣叫
50 分贝	办公室噪声
60 分贝	讨论的声音
70 分贝	街道上的交通噪声
80 分贝	真空吸尘器的声音
90 分贝	管弦乐队演奏的声音
100 分贝	链锯工作的声音
110 分贝	喷气式飞机的声音
120 分贝	听觉痛阈

一般来说，声音超过 100 分贝是令人不舒适的高声，110～1130 分贝的声压会导致人有疼痛的感觉。自然界的声音可以从几乎察觉不到的树叶沙沙声到超过 100 分贝的雷声。

当声音振动时，会产生听觉。内耳感受到声波，声音其实是在听觉皮层中产生的。听觉通道与中枢神经系统尤其是控制情绪行为的边缘系统相连通。这意味着听觉感知是通过情绪、记忆和注意力来处理的。我们接触到的所有声音都不会进入我们的意识。这是因为中枢神经系统处理通过耳朵进入的信息。如果我们认为某个声音没什么意义，我们便不会关注它。另外，一个新的、之前不为人知的声音或根据我们以往经验确认为重要的声音会引起我们的警觉。在这些情况下，我们的听觉通道会关注于倾听这个信号。当我们听到一个新的或者威胁性的声音时，我们会变得格外警觉。听觉通路旨在加强对这类声音的感知。这些被认为重要的声音可能导致焦虑、恐惧或者其他负面情绪。

We react to sounds and interpret them based on the meanings we have assigned to them. Dishes clanking in the kitchen could be a sound that makes a child feel safe but feels like annoying clatter to an adult who has trouble sleeping at night. In psychological definitions, noise is discussed using the concept of emotional annoyance that expresses the emotional impact caused by the auditory perception. In addition to the volume of the noise, its information content affects the level of annoyance. Noise with content that can be understood, such as neighbours quarrelling next door or a colleague's phone discussions in an open-plan office, are usually experienced to be more annoying than noise that contains little information. Traffic noise, for instance, has little information content. Also sudden, impulsive and tonal sounds make noise more annoying. The experience of annoyance increases if the person perceiving it experiences the sound to be unnecessary and thinks that the person causing the noise doesn't care about the wellbeing of others. If you think the noise is a risk to your health or if it is connected to fear, it annoys you more.

It has also been detected that if you are unsatisfied with other features of the environment, such as unsafety or uncleanliness of your neighbourhood, the dissatisfaction also increases the experienced annoyance of noise. The annoyance also increases if you feel like you cannot affect the noise. A soundscape based largely on nature sounds is the most functional and safest solution when trying to control the sounds in our environment. Studies conducted in different countries have shown that nature sounds, such as water flowing or the sound of the rain, rustling leaves or bird song, are soundscapes that are experienced to be positive, on average. Noise is in our environments a great problem. Traffic is one of major source of our environmental noise. Noise affects on our health quite a lot, sometimes it even kills.

It is important to prevent all kind of noise as much as possible. Shopping malls are full of non-healthy noise.Traffic noise is prevented with noise barriers in some places, but that is not enough. Acoustics is one important element of healing architecture.

我们对声音做出反应，并根据我们赋予它们的意义来解释它们。厨房里碗碟叮当作响的声音能让孩子有安全感，但是对于晚上睡眠不好的成年人则是一种烦人的噪声。从心理学定义上看，噪声是用情绪烦恼程度的概念（表达听觉感知引起的情绪影响）来讨论的。除了噪声的音量之外，噪声的信息含量也会影响情绪烦恼程度。内容可以被理解的噪声，例如隔壁邻居的争吵声、在开放式办公室里同事电话讨论声音，通常会比包含少量信息的噪声更令人讨厌。例如交通噪声几乎就没什么。此外，突然的声音、脉冲噪声和音调的声音也会更加令人烦躁。如果一个人认为某个声音是不必要的，并认为制造噪声的人并不关心他人健康时，烦恼程度也会增加。如果你认为噪声威胁你的健康或者与恐惧相关，噪声带来的烦恼度就更高。

另外还发现，如果你对环境中的其他要素不满，例如邻里空间不安全、不卫生，也会增加烦恼程度。如果你觉得你对这种噪声无能为力时，烦恼度也会增加。在试图控制环境中的声音时，主要基于大自然的声景是最有效和最安全的解决方案。不同国家的研究均表明：流水声、雨声、树叶的沙沙声或者鸟鸣声等自然界的声音，通常都是最具积极意义的声音景象体验。环境中的噪声是个大问题。交通噪声是环境噪声的主要来源之一。噪声对我们的健康影响很大，有时甚至危害生命。

尽可能地防止噪声非常重要。购物中心充斥着不健康的噪声；而有些地方用隔音屏障来防止交通噪声（图35），但这是不够的。声学设计是治愈性建筑设计的重要要素。

图35　Noise barriers Kannelmäki-Latokaski, Helsinki, Architect: Esa Piironen, 2002　赫尔辛基坎内尔姆凯－拉托卡斯基的隔音屏障，建筑师：埃萨·皮罗宁，2002

Acoustics is based on physic, but it is also connected to the vision. We need more research how those two senses are working together. Neuroscience will give us new information how our senses are working alone and together with other senses.

We are all aware of the recuperative power of sleep. A considerable body of the research documented negative effects of disputive noise on patients including raised stress, elevated blood pressure, increased heart and respiration rates, and worsened sleep. The almost standard use of hard, washable and sound reflecting surfaces creates poor acoustic conditions, causing sounds to overlap, echo and linger. The studies identified many examples of background and peak noise levels in hospitals exceeding more than twice that recommended in the international World Health Organization WHO guidelines. In multi–bed rooms, noises stemming from the presence of other patients were often seen as a major cause of sleep loss.

One study contrasted the impact on patients by alternating standard ceiling tiles with high–quality sound absorbing tiles in an intensive coronary care unit. When the sound–absorbing ceiling tiles were in place, patients slept better, were less stressed and reported that nurses gave them better care.

Air

Climate change is among the greatest challenges of humanity. Our outdoor air is at risk of being polluted at a quickening pace because of different reasons. Burning coal is one of the most significant causes of climate change. Emissions from factories and traffic also pollute the air significantly. More and more large cities have begun to be covered in smog. This was the case in London decades ago until they stopped burning coal. This is a significant matter and a responsibility of the entire society. Bill Gates has warned that capitalism cannot clean the climate. It requires a joint effort of all citizens.

Clean indoor air, on the other hand, is an issue to be taken into account by architects. There has been plenty of discussion around indoor air as modern

声学基于物理学，但也与视觉相关。我们需要做更多的关于视觉和听觉如何协同工作的研究。神经科学将为我们提供新的信息，告诉我们各感官是如何单独工作及协同工作的。

我们都意识到了睡眠对人的恢复功效。相当多的研究记录了争吵性噪声对病人的负面影响，包括压力增加、血压升高、心率和呼吸率加快、睡眠质量下降等。几乎成为标准使用的硬质的、可清洗的声音反射面材创造了糟糕的声学环境，导致环境中声音的重叠、回声和停留。研究发现，许多医院的背景噪声和峰值噪声水平远超世界卫生组织建议水平的两倍以上。在多床病房中，其他病人的噪声是造成病人失眠的主要原因。

有一项研究对比了在冠心病重症监护室使用标准天花板和高质量吸音天花板对患者的影响。当使用高质量吸音天花板时，患者睡得更好，压力更小，护士对他们的照料也更好。

空气

气候变化是人类面临的最大挑战之一。由于种种原因，我们的户外空气正面临着被加速污染的危险。燃煤是造成气候变化的重要原因之一。交通尾气和工厂的排放物也是污染空气的重要因素。越来越多的大城市被雾霾所笼罩。几十年前的伦敦就是这种情况，直到他们停止了烧煤。这是一件大事，也是全社会的责任。比尔·盖茨（Bill Gates）曾警告过人们：资本主义并不能净化气候，这需要全体公民的共同努力。

另外，干净的室内空气是建筑师需要考虑的一个重要问题。随着现代建筑结构从自然通风转变为机械通风，关于室内空气的相关探讨越来越多。与此同时，建筑结构和建筑材料没有跟上，这导致越来越多的

construction transferred from natural ventilation to mechanical ventilation. At the same time, the structures and materials of buildings didn't keep up, which resulted in more and more houses being infested with mold. In the northern climate, where temperatures vary drastically from one season to another, humidity remains inside the walls. Over time, the humidity will create mold that is harmful or even perilous to human health.

The issue can be fixed with good planning but it takes years. We also need better cooperation in planning and design between the architect, structural engineer and ventilation engineer. This also affects education.

Indoor air has always more or less impurities that we breathe in daily at home or work. The impurities can derive from, for instance, construction or decoration materials, furniture, devices, etc. Detergents, pets, cooking or perfumes add more irritants to our indoor air. Particles can also travel from the outside to indoors, for example, traffic emissions, pollen and mold spores. Most common symptoms caused by poor–quality indoor air include irritation of eyes, respiratory tracts and mucous membranes, skin irregularities (dryness, itching, redness, rashes) and symptoms of fatigue (headache, dizziness).

A very general emission of chemical impurities is formaldehyde that can be released from chipboard, laminates, parquets, glues and textiles. Plants' ability to remove this very common toxin from our alveolar air has been studied for long. Already in 1980, NASA's space research centre found out in their study that house plants can remove vaporizable organic matter in a closed test chamber.

Luscious and healthy plants purify air the best. Plants that purify indoor air absorb impurities and the microbes in their roots decompose toxins to be used as nutrition of the plants and micro–organisms.

房屋易于滋生霉菌。北方的气候，气温因季节不同而变化很大，潮气积聚在墙体内。经年累月，潮气将滋生对人体有害甚至威胁人类生命的霉菌。

这个问题可以通过良好的规划来解决，但需要数年的时间。同时也需要建筑师、结构工程师和暖通工程师在规划和设计上的良好合作。这反过来也会影响专业教育。

我们在家或在办公室日常呼吸的室内空气中，或多或少都会有些杂质。这些杂质来自建筑、装饰材料、家具、设备等，而洗涤剂、宠物、烹饪或者香水也会增加室内空气中的刺激物。还有些微粒是从室外进入室内的，比如交通尾气、花粉、霉菌孢子等。室内空气质量差导致的常见症状包括眼睛、呼吸道和黏膜发炎，皮肤问题（干燥、发痒、发红、皮疹）以及疲劳（头疼、眩晕）。

一种非常普遍的化学杂质是甲醛，它会从刨花板、层压板、胶合板以及纺织品中释放。植物清除我们体内甲醛的能力已经被研究了很长时间。早在 20 世纪 80 年代，美国国家航空航天局空间研究中心就发现，植物可以在密闭的实验空间中除去可蒸发的有机物（图 36、图 37）。

芬芳健康的植物最有利于净化空气。植物吸收杂质来净化室内空气，植物根部的微生物将毒素分解为植物和微生物可吸收的营养（图 38～图 40）。

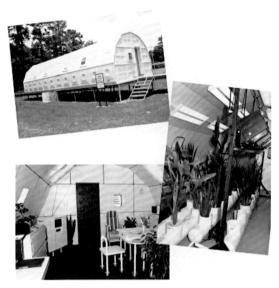

图 36　Nasa module　美国国家宇航局实验空间

图 37　Mars habitat, Norman Foster
诺曼·福斯特设计的火星栖息地

Plants that purify indoor air include:

Golden cane palm, Chinese evergreens, bird's-nest fern, spider plant, flaming Katy, peace lilies, parlous palm, rubber fig, dumb cane, Ceylon creeper, weeping fig, lacy tree philodendron, snake plant, areca palm and many others.

According to the NASA's study, at least one plant per approx. 9 square meters is required to purify the indoor air of homes and offices using house plants. Plants absorb carbon dioxide and release oxygen indoors.

图 38　Country Garden, Foshan, China　碧桂园立体绿化，中国佛山（一）

图 39　Country Garden, Foshan, China　碧桂园立体绿化，中国佛山（二）

图 40　Wall of Nature in the office 办公空间中的绿墙

净化室内空气的常见植物包括：

金藤棕榈、中国常青树、燕窝蕨、蜘蛛草、长寿花、和平百合、棕榈树、橡胶树、哑巴藤、锡兰爬山虎、垂叶榕、花边树、蛇藤、槟榔树等。

根据美国国家航空航天局的研究，使用植物净化家庭和办公室室内空气，每9平方米至少需要一株植物。植物在室内吸收二氧化碳释放氧气。

Nature

Nature is the teacher of design, its
source of inspiration and store of
materials.

> —Tapio Periäinen

It has been known for a long time that the natural environment is closely related
with health and its ambient environmental conditions affect human health.
However, there is very little evidence to suggest that the physical aspects of
built environment can affect human health. The relationship between natural
environment and health is clearly explained in the Hippocratic treatise. The treatise
basically emphasizes that climatic and geographical factors have strong influence
on human health. Most physicians accept that illness emerges due to the action of
the three factors: disposition (e.g. genetic), stress (e.g. exhaustion) and agent (e.g.
pathogen). Whether or not we trace causes of the illness materially, psychologically
or spiritually, environment has a significant role in all parts of the process.

Through evolution, we are part of nature. Architecture was created to shield us
from nature. However, over history, we have become alienated from nature.

Global urbanisation is progressing rapidly. As a response, humans should strive
to get closer to nature, and architecture should aid us with our relationship with
nature. In addition to green and trees, this also applies to the sky, water and air.
Humans feel comfortable close to nature. Forests have a calming effect on most
Finns. Nature empowers people's lives, and can, at best, cure people. It has
been proved that the colour green alone calms us down. We build space
for life. Architecture also helps us understand life in the time the building
was created in.

Already the columns of Greek temples showed characteristics of mimicking
nature. Nature is the role model of architecture in many fashions; both must

自然

自然是设计的老师，是
灵感之源，是材料宝库。
——塔皮奥佩·里伊宁

人们很久之前就认识到自然环境与人类健康息息相关，其周围的环境影响着人类的健康。然而，很少有证据表明建筑环境会影响人类健康。希波克拉底的论著清楚地解释了自然环境与人健康的关系。他的论著主要强调了气候和地理因素对人类健康的巨大影响。许多内科医生认为疾病的出现是由于三个因素的作用：性情（如遗传）、压力（如疲劳）和作用物（如病原体）。无论我们是否从物质上、心理上或精神上追踪疾病的原因，环境在这个过程的各个环节都起着重要的作用。

在整个进化过程中，人类是自然的一部分。建筑用来庇护人类，免于自然的侵害而随着历史的发展，我们已经和自然疏离。

全球城市化进程正在迅速推进。作为城市化的回应，人类应该努力接近自然，建筑应该帮助人类处理好人与自然的关系。除了绿色和树木，天空、水和空气也同是自然，接近它们时人们会感到舒服。森林让我们大多数芬兰人感到宁静。大自然赋予人们生命的力量，也能在最大程度上治愈人们。已有研究证明，只有绿色能令我们平静。我们为人们的生活建造空间，建筑有助于我们了解其所建造时代的生活。

希腊神庙的柱子已经显示出模仿自然的特征。自然在很多方面都是建筑的典范，建筑与自然都必须遵循重力原则。自然界的发展常常激发

be in agreement with gravity. Nature's growth has often inspired architecture. Natural materials have through time controlled the methods that are possible for architecture. Examples of natural forms and inspiration from nature in architecture have been part of an architect's work for long.

图 41　Finnish Landscape　芬兰景观

建筑设计的灵感。随着时间的推移，自然材料掌控着建筑设计可能的
方式。长期以来，建筑中的自然形式和来自自然的灵感一直是建筑师
工作的一部分（图41、图42）。

图42 Ecological House, Architect: Bruno Erat　生态房屋，建筑师：布鲁诺·埃拉特

However, we must take inspiration from nature further. Nature consists of cells and particles smaller than cells. Every snowflake is different because of nature's standardisation. Similar kind of thinking should be applied to design and architecture. Now that the humanity has figured out DNA and many other natural phenomena, we should sternly focus on developing that information in order to create new design paradigms.

That could be our way towards so–called healing architecture.

Many Finns view nature and forest as home. The reason might be that we only recently, compared to many other peoples, moved from the forest to cities, or the cause could be something else entirely. This hasn't been studied much.

But it has been studied that walking in a forest can clearly lower your blood pressure.

Nature rooms have been started to be constructed in offices and elsewhere where you can take a 45–minute break between work tasks. A nature room can include nature videos and nature sounds. Studies have shown that people's blood pressure lowers in these kinds of premises and they promote relaxation.

Color

Color is an integral element of our world, not just in the natural environment but also in the man–made architectural environment. Color always played a role in the human evolutionary process. The environment and its color are perceived, and the brain processes and judges what it perceives on an objective and subjective basis. Psychological influence, communication, information, and effects on the psyche are aspects of our perceptual judgment processes. Hence, the goals of color design in an architectural space are not relegated to decoration alone.

Especially in the last eleven decades, empirical observations and scientific studies have proven that human–environment–reaction in the architectural environment

然而，我们必须进一步从自然中汲取灵感。自然由细胞和比细胞更小的粒子组成。自然法则使每一片雪花都不一样。类似的思维应当运用在设计与建筑中。现在人类已经发现了 DNA 和其他很多自然现象，我们应当坚定地关注这些信息的发展以创造出新的设计范式。

那可能就是我们通往所谓"治愈性建筑"之路。

许多芬兰人都视自然和森林为家。可能是因为与其他国家的人们相比，我们芬兰人才从森林走向城市，或者也可能完全是其他原因。这方面没有太多的研究。

但研究表明，在森林里散步能明显降低血压。

人们已经开始在办公室建造自然式的空间，你可以在工作任务之余在这一空间中休息 45 分钟。一个自然式空间可以包括自然界的影像和自然界的声音。研究表明，在这样的空间中，人们的血压会降低，也会有助于人们的放松。

色彩

色彩是我们这个世界不可或缺的元素，不仅在自然环境中是如此，在人造建筑环境中也是如此。色彩在人类的进化过程中始终扮演着重要角色。环境及其色彩能被人们所感知，大脑在客观和主观结合的基础上判断和处理所感知的事物。我们感知判断的过程包括：心理的影响、交流、信息以及对心灵的影响等。因此，建筑中色彩设计的目的不仅仅是简单的装饰。

尤其是在过去的 110 年中，经验的观察和科学研究均证实建筑环境中的"人—环境—反应"在很大程度上基于对色彩的感知。这些研究包括心

is to a large percentage based on the sensory perception of color. These studie include the disciplines of psychology, architectural psychology, color psychology, neuropsychology, visual ergonomics, psychosomatics, and so forth. In short, it confirms that human response to color is total—it influences us psychologically and physiologically.

Color is a sensory perception, and as any sensory perception, it has effects that are symbolic, associative, synesthetic, and emotional. This sel–evident logic has been proven by scientific investigation. Because the body and mind are one entity, neuropsychological aspects, psychosomatic effects, visual ergonomics, and color's psychological effects are the components of color ergonomics. These being design goal considerations that demand adherence to protect human psychological and physiological well–being within their man–made environment. The color designer has the task of knowing how the reception of visual stimulation, its processing and evoked responses in conjunction with hormonal system, produces the best possibilities for the welfare of human beings. This is of utmost importance in various environments, such as medical and psychiatric facilities, offices, industrial and production plants, educational facilities, homes for the elderly, correctional facilities, and so forth. Each within themselves having different task and function areas.

Color has not always been so detached from architectural design. Historically, the artist's profession encompassed all, but not exclusively: painting, sculpture, and architecture. Color was used lavishly in architecture, because of the desire to glorify gods , kings or the building itself. The thought–to–be bare and neutral stone temples of ancient Greece have been proven to have been richly painted with deep jewel–toned pigments. The cathedrals of medieval Europe were also painted, as well as the palaces and temples of China, which were filled with color symbolism.

理学、建筑心理学、色彩心理学、神经心理学、视觉工效学、心身医学等学科（图43、图44）。简而言之，研究证实了人们对于色彩的反应是全面的——它影响了我们的心理和生理。

色彩是一种感观感受，如同其他感观感受一样具有象征的、联想的、联觉的和情感的效应。这一不言自明的逻辑已被科学研究所证明。身心是一个整体，神经心理学、心身医学、视觉工效学和色彩心理效应都是色彩工效学的组成部分。这些都是设计目标要考虑的，要求坚持在人造的环境中保护人类的心理和生理健康。色彩设计师的任务是了解视觉刺激中人的接受过程，这一过程的处理，以及与荷尔蒙系统相关的诱发反应，为人类福祉提供了最好的可能性。这在各种环境中至关重要，例如医疗和精神疾病设施、办公室、工业和生产工厂、教育设施、老人之家和监狱设施等。每一项都有不同的任务和职能领域。

一直以来，色彩并非都与建筑设计割裂。历史上艺术家的职业范畴涵盖广泛，不限于绘画、雕塑和建筑。建筑中大量使用色彩，是因为人们想要赞美上帝或者国王或者建筑本身。显得光秃秃的和中性的古希腊石头庙宇，已经被证明曾是用深宝石色调的颜料来上色的。欧洲中世纪的教堂都曾刷以颜色，中国的宫殿和庙宇也用了充满象征意义的颜色粉刷。

伊顿12色相环是由近代瑞士色彩学大师约翰内斯·伊顿（Johannes Itten, 1888-1967）设计，特点是由颜料的三原色混合叠加而成。

伊顿12色相环

一次色（原色）：
在美术上，将红、　、蓝称为颜料的三原色或一次色。

二次色（间色）：
通过两种不同比例原色进行混合所得到的颜色为二次色，二次色又叫做间色。

三次色（复色）：
用任何两个间色或三个原色相混合而产生出来的颜色为三次色（复色），包括了除原色和间色以外的所有颜色。

图43　Colors by Itten　伊顿色相环

One of the most striking results concerning color connotations and color mood associations is its consistency cross–culturally from one individual to another and group to group. The great number of studies comparing human subjects worldwide, such as men to women, children to adults, laymen to architects, and even monkeys to humans show that color is an international visual language understood by all.

The impression of a color and the message it conveys is of utmost importance in creating the psychological mood or ambiance that supports the function of a space.

During the 1960s the psychological and even physiological impact of color started to become a consideration. Following in Goethe's footsteps, Faber Birren (1900–1988) was one of the first people to do extensive research on the human perception of and response to color. He wrote over 20 books and 200 articles on the topic. Today contemporaries such as Frank and Rudolf Manke and Carlton Wagner are picking up where Faber Birren had left off. This section of human color response will first cover the functions color can have in our everyday environments, then it will expand on the different levels of experience we have, and finally it will discuss the primary and secondary hues and their specific effects on people and in spaces.

在颜色内涵和颜色情绪关联方面，最引人注目的结果之一是，它在从一个人到另一个人、从一个群体到另一个群体的跨文化上，具有一致性。世界范围内以人类为对象的大量对比性研究表明，对于男人和女人，孩子和成人，外行和建筑师，甚至猴子和人类，颜色都是一种能直观理解的通用视觉语言。

人们对某一种颜色的印象及其所传递的信息，在创造心理情绪和营造氛围以便协助提升空间功能上，起着至关重要的作用（图45）。

20世纪60年代，人们开始关注色彩对人的心理甚至生理的影响。继歌德之后，费伯·比伦（1900–1988）是广泛研究人类对色彩感知和反应的第一人。关于这个主题，他撰写了20多部著作和200多篇论文。当代建筑师弗兰克、鲁道夫·曼克及卡尔顿·瓦格纳，在费伯·比伦的研究基础上，继续推进相关研究。人类对色彩的反应，首先涵盖的是色彩在日常生活中的作用，其次延伸到不同层次的体验，最后讨论的是关于主次色调及其对人们和空间的具体影响。

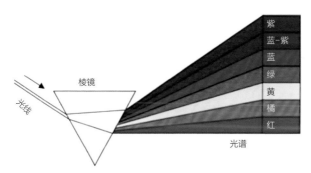

可见光谱的七种光波中，紫色波长最短，红色波长最长。对光谱颜色的仔细观察也揭示了人眼可以在它们之间检测出不同的颜色，比如红-橙之间的橘红色，蓝-绿之间的蓝绿色。

图44 Color from prism 三棱镜折射七色图

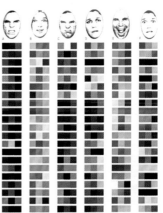

图45 Colors and moods 颜色与心情

As you can see, color has many influences in our everyday lives. We have learned to respond to certain colors in certain ways. For example, red means caution/stop/blood, but there are also reactions that are unconscious. Biological reactions to color are solely physical in nature. Instead of the obvious optical reaction to color, it is in fact a reaction to the energy of the light waves. Tests show that even if a person is blindfolded his or her pulse will noticeably increase when exposed to the color red and decrease when exposed to blue. This reaction to color is also not governed by the intellect. It is a reaction that originates out of our genetic imprinting. In some cases it might also be triggered by a former personal experience. For example, after an accident, a person might dislike the color red without consciously making connection to the color of blood.

Conscious Symbolism develops through personal experiences. There are some universal associations that are surprisingly uniform from culture to culture. Blue for example is usually associated with sky and water, yellow with sun and light, and red with blood and fire. There are also cultural influences on our experience of color. For example, in English language, if a person is said to be green he/she feels sick; in German, on the other hand, if a person is said to be green he/she is hopeful.

Almost every year there are new color trends, especially in fashion. Even though color trends are short–lived they still influence our associations. However, it is not useful for the architect to follow these color trends since they hardly ever consider psychology or visual ergonomics.

Our personal relations to color vary greatly. It is a field that the designer has hardly any control over. Generally speaking younger people prefer more saturated and primary colors where older people prefer less saturated and subdued colors. The same logic exists for extroverts and introverts.

如我们所见，色彩对日常生活有很多影响。我们已经学会了对特定的色彩做出特定的反应。例如，红色代表注意、停止、血等，但也有一些反应是无意识的。对颜色的生物性反应在本质上完全是物理性的——不是对颜色的光学反应，而是对光波能量的反应。测试显示，即使一个人被蒙住眼睛，他／她的脉搏跳动会在红色环境中显著增加，在蓝色环境中显著下降。这种对颜色的反应也不受智力的支配。它是一种源自我们基因印记的反应。在某些情况下，这也可能是由以前的个人经历触发的。例如，某个人在一次事故后，即使在没有特意将红色与血液联系，可能依旧不喜欢红色。

有意识的象征主义是通过个人经历发展起来的。不同文化中一些普遍的关联性有着令人惊讶的一致性。例如蓝色常常与水、天空相联系，黄色与太阳、阳光相联系，红色则与血、火相联系。我们对于色彩的体验也受文化的影响。例如，在英语中，如果说一个人是绿色的，意思是他病了；而在德语中，如果说一个人是绿色的，说明他充满了希望。

几乎每年都有新的色彩趋势，尤其在时尚界。即使色彩趋势是短暂的，它仍然会影响我们的联想。然而，由于色彩趋势几乎从不考虑心理学和视觉工效学，建筑师去跟随色彩趋势便毫无用处。

我们每个人和色彩的关系差别非常大。这是一个设计师几乎无法控制的领域。通常来讲，年轻人更喜欢饱和色和原色，而老年人更喜欢不饱和色和柔和的颜色。同样的逻辑也适用于外向型性格和内向型性格的人群。

Form

In psychology there is certain facts that most people may know by instinctively. The so–called takete and malumma forms in our mind respect to the pointed and rounded forms. The former form is not so inviting and the latter quite caressing. These forms we see also in architecture. Daniel Libeskind designed Denver Art Museum annex resembles that takete form.

Architectural trends come and go. Modernism brought new ideas to our way to build. It had some good ideas, but it forgot human being, history and tradition. After that we had post–modernism that wanted to through all ideas away starting to imitate classical motifs. Then came late modernism that tried to save what is to be saved. Wow–architecture is today still among us, but I hope science can help architects to face the facts and design sustainable, health buildings for human beings.

形式

在心理学中，有些事实是大多数人凭直觉获取的。所谓的塔克特和玛鲁玛形式是指尖锐的形式和圆润的形式。前一种形式不那么吸引人，而后一种形式则很讨人喜欢。我们在建筑中也能看到这些形式。丹尼尔·里伯斯金设计的丹佛艺术博物馆附属建筑是塔克特（尖锐形式）形式的代表（图46）。

建筑潮流来来去去，现代主义带来了新的建造想法。有些想法很好，但它忘记了人类本身、历史以及传统。之后我们有后现代主义，想要通过所有的想法去模仿古典主义；再后是试图去拯救需要被拯救之物的晚期现代主义。今天仍有很多奇怪的建筑围绕着我们，但我希望科学可以帮助建筑师面对现实，为人类设计可持续且健康的建筑。

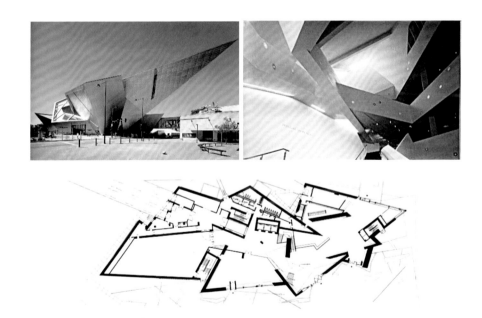

图46　Denver Art Museum Annex, Daniel Libeskind　丹佛艺术博物馆附属建筑，丹尼尔·里伯斯金

Scale and Proportion

Humans interact with their environments based on their physical dimensions, capacities and limits. The field of anthropometrics (human measurement) has unanswered questions, but it is still true that human physical characteristics are fairly predictable and objectively measurable. Buildings scaled to human physical capabilities have steps, doorways, railings, work surfaces, seating, shelves, fixtures, walking distances, and other features that fit well to the average person.

Humans also interact with their environment based on their sensory capabilities. The fields of human perception systems, like perceptual phycology and cognitive psychology, are not exact sciences, because human information processing is not a purely physical act, and because perception is affected by cultural factors, personal preferences, experiences, and expectations. So human scale in architecture can also describe buildings with sightlines, acoustic properties, task lighting, ambient lighting, and spatial grammar that fit well with human senses. However, one important caveat is that human perceptions are always going to be less predictable and less measurable than physical dimensions.

When architects talk about scale and proportion they are usually talking about how the individual parts of the project relate to each other, how the project relates to the size of the human body and how the project relates to its contextual scale. A project that's perceptually successful is a project that takes these factors into account.

To ensure a sensory consistency, individual project components such as rooms, wall finishes, ceiling shapes and finishes, windows, doors, built–ins, etc. should have a compatible scale from one to another.

尺度与比例

人类与环境的互动基于人们的身体尺寸、能力和限制。人体测量学领域尚有许多未知领域，但人类的身体特征是基本可预测和客观可衡量的。根据人体尺度设置的建筑结构有台阶、入口、栏杆、工作台面、座椅、搁板、固定装置等，设计中应适合于普通人的身体特征。

人类和环境的互动也基于人们的感官能力。人类的感知系统领域，如知觉心理学和认知心理学，并非严谨意义上的科学，因为人类的信息处理并不是一种纯粹的物理行为，而且感知会受文化因素、个人偏好、个体经验和个体期望的影响。因此建筑中的人体尺度也可以用符合人类感官的视线、声学特性、功能照明、环境照明以及空间语言等来描述。然而，需要注意的是人类的感知总是比物理维度更难以预测和量化。

当建筑师谈到尺度与比例，他们通常谈的是建筑各个部分如何相互关联，建筑如何适合人体的尺度，以及建筑如何与周边环境尺度相联系。一个被认为成功的项目，都会将以上这些因素都考虑在内。

为了确保感官上的一致性，项目中的各个组成部分如房间、墙面装饰、天花板形式与饰面、窗户、门和内置物等都应有一个相对应的尺度。

Architects understand that the starting point for our perception of something is the size of our own bodies. Whether or not a room is large or small or somewhere in between has a direct correlation to how we understand that room in relation to our size. So a room that's overly large or overly small can make us uncomfortable. A very tall, high ceilinged, room that's small in area can make us feel as if we are in a pit. So giving a space "human proportions" increases the likelihood that we'll find the space comfortable.

There are proportional systems that can help an architect develop a design that is responsive to the human figure. The most used is that system based on the golden ratio. This proportional system is found in nature (most famously the Nautilus shell), music and we come into contact with it every day when we turn on a light (the dimensions of the typical wall switch cover plate correspond to the golden ratio).

We are building our buildings mostly for human beings. Sometimes we build monumental buildings that I hope is a vanishing trend. We should think more children, disabled and old people when designing our buildings. Of course the right scale is always our goal.

Ergonomics

Ergonomics (or human factors) is the scientific discipline concerned with the understanding of interactions among humans and other elements of a system, and the profession that applies theory, principles, data and methods to design to optimize human well–being and overall system performance.

Proper ergonomic design is necessary to prevent repetitive strain injuries and other musculoskeletal disorders, which can develop over time and can lead to

建筑师应理解我们感知事物的出发点是我们自身的身体尺度。一个房间是大还是小，或者介于两者之间，与我们如何理解房间与自身身体尺度的关系直接相关。因此，一个房间过大或过小，都会让我们感到不适。一个天花板很高、面积很小的房间，会让我们觉得身陷坑中。因此赋予空间"人体尺度"增加了我们发现空间舒适度的可能性。

图47 Scale and proportion 尺度与比例

一些尺度系统可以帮助建筑师做出符合人体尺度数据的设计。最常用的是基于黄金分割比的系统。该尺度系统存在于自然界（最著名的是鹦鹉螺贝壳，见图47）、音乐，以及我们日常生活每天开灯所接触的开关面板（墙壁开关盖板尺寸通常对应黄金分割比）。

我们主要为人类建造建筑。有时我们会建造巨大的建筑，我希望这样的建筑能逐步消失。我们在设计和建造时应更多地考虑儿童、残障人士和老年人。当然，正确的尺度一直都是我们的目标。

人体工程学

人体工程学（人机工程学）是一门研究人类和系统中其他元素相互作用的学科，是应用理论、原则、数据和方法进行设计，以优化人类福祉和整体系统性能的专业。

为了防止随着时间的推移而发展并会导致长期残疾的重复性劳损和其他肌肉骨骼疾病，正确的人体工

long–term disability. Human factors and ergonomics is concerned with the "fit" between the user, equipment, and environment.

Ergonomics comprise three main fields of research: physical, cognitive and organizational ergonomics. There are many specializations within these broad categories. Specializations in the field of physical ergonomics may include visual ergonomics. Specializations within the field of cognitive ergonomics may include usability, human–computer interaction, and user experience engineering.

Environmental ergonomics is concerned with human interaction with the environment as characterized by climat, temperature, pressure, vibration, light.

Physical ergonomics is concerned with human anatomy, and some of the anthropometric, physiological and bio mechanical characteristics as relate to physical activity. Physical ergonomic principles have been widely used in the design of both consumer and industrial products. Risk factors such as localized mechanical pressures, force and posture in a sedentary office environment lead to injuries attributed to an occupational environment.

Ergonomics is one of the most important objective when designing environment and furniture. In architecture we need ergonometrics practices to make better design solutions. Acoustic design is one way to design with ergonomics.

Stairs in a building can be designed so that they are easy to rise without breathing. You have to know the right rise and advance in stairs to design that.

程学设计是非常必要的。人体工程学关注使用者、设备和环境之间的相互匹配。

人体工程学包括三个主要研究领域：身体、认知和组织工效学。这三个领域有许多专业类别。身体功效学的专业类别可以包括视觉工效学。认知功效学包括可用性、人机交互和用户体验工程等。

环境工效学研究人与环境，诸如气候、温度、气压、振动、光等的相互作用。

身体功效学涉及人体解剖学，以及与身体活动相关的人体测量、生理学和生物力学特性。身体功效学原理已被广泛运用于消费产品和工业产品的设计中（图48）。在久坐的办公室环境中，局部的机械压力、受力和姿势等风险因素会导致职业环境中的伤害。

人体工程学是环境设计和家具设计中最重要的目标之一。在建筑设计中，我们需要利用人体工程学的设计实践来做出更好的设计解决方案。声学设计是结合人体工程学进行设计的方式之一。

建筑中的楼梯可以设计得易于爬升而不令人气喘吁吁。设计楼梯时，设计师必须知道正确的高度和上升方式等。

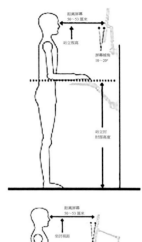

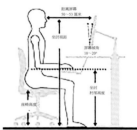

图 48 Ergonometrics measures
人体工程学尺寸

Materials

We see daily a lot of building materials in our environment, at least if we live in a city.

Sometimes there is a visual chaos. We also touch many materials daily.

Materials belong to our life and buildings.

Architects choose materials designing buildings. Do we know enough materials?

How the materials have an effect on human life? Are materials healing our environment or are we building more sick buildings. That means also more sick people.

We have to remember that architects are designing for human beings. Materials are actually made of stardust in the beginning. We have to know how the neutrons, protons and atoms are working in universe.80% of material in the universe is dark matter, that we do not see or know too much.

The material is an accumulation of particles which all follow their own rules of physics. Architecture cannot exist without the material. The relationship between an idea and the material is an object of ongoing investigation. Volumes have been written about materials. Much of this literature focuses on the physical qualities of materials. This knowledge is of great importance to builders and designers. There is, however, less research on the impact of materials on people's wellbeing. The general principle goes that the right materials should be used in the right place in a building. The way materials interact depends on the qualities of these materials. In architecture, however, the idea is more important than the material.

And yet it is the synthesis of the material and the idea that generates truly high–quality architecture.

材料

生活在城市中，我们每天都会在周围环境中见到大量的建筑材料。

有时候这些材料会给人以视觉上的混乱。

我们每天也会触摸到许多材料，材料属于我们生活和建筑物的一部分。

建筑师选择不同的材料来设计建筑物。我们是否知道足够多的材料？

材料如何影响人类的生活？材料对我们的环境有治愈性，还是我们在建造一些不健康的建筑？不健康的建筑意味会带给人更多的不健康。

建筑师必须记住自己是在为人类设计建筑。材料最初由星尘组成。我们必须要探知中子、质子和原子在宇宙中是如何工作的。宇宙中 80% 的物质是暗物质，我们看不到或不知道的太多。

材料是粒子的累积，它们都遵循各自的物理规律。没有材料，建筑就不可能存在。设计想法和材料的关系持续存在。关于材料的著作已经有许多。这些文献多集中关注于材料的物理特性。这些知识对于建造和设计师来说非常重要。然而，很少有关于建筑材料对人类健康影响的研究。总的原则是：正确的材料用在正确的建筑部位。材料间的相互作用方式取决于这些材料的质量。在建筑设计中，设计理念比材料更重要。

然而，材料和理念的结合才能产生真正高质量的建筑。

Buildings are built of different materials. An architect has to choose most suitable materials to the buildings he is designing; the best material to a certain point of a building.

A building consists normally the roof, the walls, the floor and the base or basement. For all these parts, there are lot of possibilities to choose materials. Depending where an earth you are working, materials to choose for these parts of a building differ a lot.

Exterior materials of a building must last as long as possible in a normal use, but weathers are very different in various parts of the world. Temporary buildings have their own rules. Also the interior of a building must have habitable materials; healing materials if possible. An architect has to know how materials behave in our environment and how the materials can be put together as well.

According to recent studies, the use of wood in interiors has a surprisingly positive effect on people. The research says that wood is a material which underpins good health and supports recovery.

Using wood can influence mood and stress levels. The studies show that people react favourably to wood, both physiologically and psychologically.

Wood finishes make rooms feel warmer and more cosy, and induce a feeling of calmness.

Wood seems to have the ability to regulate the body's stress levels. In a comparison of different work spaces, stress levels measured as the electrical conductivity of the skin were lowest in rooms with wooden furniture. No corresponding calming effect was observed in rooms with white furniture and indoor plants.

Touching wooden surfaces gives a soft, safe, natural feeling. On the other hand, touching stainless steel, cold plastic, or aluminum at room temperature causes an

建筑由许多不同的材料建成。建筑师必须为他设计的建筑选择最合适的材料，即某种意义上对这个建筑来说是最好的材料。

一个建筑通常由屋顶、墙面、地板和地基或地下室组成。对于这些组成部分，每个部分都有很多的选材可能。根据所在地理区位不同，建筑材料的选择会有很大不同。

建筑表皮材料要保证正常使用条件下的尽可能耐用，但是世界各地的气候条件差别很大。临时性建筑有自己的原则。此外，建筑物的内饰材料必须是适宜居住的材料，如果有可能，最好是治愈性材料。建筑师必须懂得建筑材料如何作用于我们的环境，以及如何将材料组合在一起。

根据新近的研究，室内设计中使用木材，有令人惊讶的积极影响。这些研究表明木材对人类的健康和康复有巨大的支持作用。

使用木材会影响人的心情和压力水平。研究表明，无论生理上还是心理上，人们都对木材表现出了喜爱之情。

木饰面让房间感觉更为温暖和舒适，同时令人感到平静。

木材似乎有能力调节身体的压力水平。对比不同的工作空间，用皮肤导电率测量压力水平，在木质家具的房间中，压力水平最低。在白色家具和有室内植物的房间中，却并没有发现相应的安抚效应。

触摸木质表面给人柔软、安全和自然的感受。而在室温条件下触摸不锈钢、低温塑料或者铝材则会导致身体血压上升。触摸木质表面没有

increase in blood pressure. The studies did not observe a similar reaction when touching wooden surfaces.

The research indicates that the positive effects cannot be achieved by using imitation wood. Physiological measurements show that quality of sleep and recovery from stressful situations were better in a room finished in wood than in a room finished in imitation wood.

Positive psychological effects have also been observed in schools. Stress peaks measured by pulse variations in classrooms finished in solid wood die away soon after pupils arrive at school, whereas in control classrooms, moderately stressed conditions continue all day long. Pupils' experiences of stress and feelings of tiredness and lack of achievement were less in wood–finished classrooms than in normal classrooms.

The effect of using wood in interiors also appears to extend to people's behaviour and social observation. In commercial spaces where wood products are used, visitors had more favourable first impressions of staff than in spaces without wood.

One interesting observation is related to housing for the elderly. When wooden materials and wood surfaces began to be used in housing for the elderly, the interaction between them and their interest in the environment increased.

Wood is also anti–bacterial. It has been proved to prevent the growth of dangerous microbes Consequently, wood is used in saunas, washrooms and kitchens.

It has long been known that wood surfaces can affect acoustics and indoor air quality. Traditionally, the acoustic properties of wood have been put to use in instruments, lecture rooms and concert halls.

Furthermore, wood has the ability to absorb and release moisture, i.e. even out changes in the humidity of indoor air. Steady humidity improves the quality of

发现相似反应。

该研究表明，使用仿木材质无法获得良好的效果。生理上的数据测量显示，木饰面房间中人的睡眠质量和压力恢复状况比仿木饰面房间中的要好。

木饰面房间产生的积极生理效应在诸多学校空间研究数据中也被发现与证实。在实木装饰的教室内测试到校学生的脉搏，得到的压力峰值在学生到达教室后不久就消失了。反之，在相应的测试教室空间中，一定的压力情况会持续一整天。与普通教室相比，在木质饰面教室中，学生的压力感、疲劳感和缺乏成就感的体验较少。

室内使用木材料的这些效应也延伸体现在人们的行为方式和社会调查中。使用木材的商业空间相较于非木质饰面商业空间，来访者对于该商业空间中的服务人员的第一印象会更好一些。

一个关于老年人住所有趣的观察发现：当木材和木饰面用到老年人住所后，老年人之间的交流和他们对于周边环境的兴趣提高了。

木材具有抗菌性，木材已被证实能防止危险微生物的滋生，因此常使用在桑拿房、卫生间和厨房中。

很久之前已被证实，木材具有隔音和改善室内空气质量的效应。通常，木材的隔音作用使得木材常被用于乐器房、报告厅和音乐厅。

此外，木材能够吸收空气中的水分，调节室内空气湿度。稳定的室内空气湿度有利于提高室内空气质量，降低通风设备的使用需求，从而

indoor air and reduces the need for ventilation, which in turn affects the energy–efficiency of the building.

The effects of wood surfaces on the body have been studied in Norway, Austria, Japan and Canada, as well as in Finnish research institutions. The reasons for these positive effects are not yet known, but they emphasize much of the traditional wisdom attached to wood.

The antibacterial qualities of wood are probably still a fairly unknown territory. However, studies have shown, for example, that the more precious the metal used in door handles, the fewer bacteria they spread.

Research has been carried out on the role of plants in so–called healing architecture, clearly showing that plants help improve the quality of air both indoors and outdoors.

When building with steel, it is crucial to understand the fundamentals of the material's behaviour.

In warm indoor spaces, hand rests, door handles, water taps and other similar items are often metal. Brass handles leave a metallic smell on the hands. Medical research has discovered that by using more noble metals in door handles and water taps (e.g. copper), the amount of bacterial build–up can be dramatically reduced in places such as hospitals and schools. This would have a positive effect on public health.

提高建筑节能性。

挪威、澳大利亚、日本、加拿大以及芬兰的相关研究机构都研究了木材表皮对人体的影响，木材表皮的这些正面作用的产生原因尚不清楚，但强调木材使用凝聚着许多传统智慧。

木材的抗菌性可能仍是一个未知领域。然而，已有研究表明，例如，门把手使用的金属材料越少，细菌的传播越少。

关于植物在治愈性建筑中的作用的研究清楚地表明植物有助于改善室内和室外的空气质量。

用钢材做建筑材料，至关重要的是了解钢材的性能。

在温暖的室内空间，扶手、门把手、自来水龙头和其他一些相似的配件常常用金属。黄铜把手会使人的手上留下一股金属味道。医学研究表明，在医院、学校这类公共建筑中，门把手、水龙头等多用贵金属材料（例如铜）可以大大减少细菌聚集的数量。这对公众健康有积极作用（图49）。

图49　Bronze door handle, Chapel of Holy Cross, Turku.　青铜门把手，圣十字教堂，图尔库

Natural light is crucial for people's wellbeing. Even in the north, where daylight is scarce in the winter, all available natural light needs to be captured. Transparency enables an illusion, an integral part of architecture.

Electrically heated glass is practical in spaces where the chill radiating through large window panes would make sitting nearby uncomfortable. Thin, transparent metal films sandwiched between two panes conducts electricity providing heating.

A similar method is used for glass that can be darkened with the press of a button.

Glass may be tempered to better withstand impact.

Safety glass is made by adding a plastic film in between glass panes. If the glass sustains an impact, the glass may break but the plastic film keeps the splinters together. Tempered glass breaks into tiny rubble.

Atmosphere

In his book *Atmospheres*, architect Peter Zumthor discusses the essence of architecture's atmosphere. It is one of the key central topics in studying and teaching architecture that must be taken into account when planning our surroundings.

Ther atmosphere is a very subjective phenomenon where different people's different susceptibility to surroundings comes forth.

Some people focus on large things whereas tiny details convey the meaning of the surroundings for others. All that is included in the concept of atmosphere is under constant scrutiny. Taking all senses into account in the atmosphere is key. For different people, the meaning of different senses is very subjective. Architects often emphasise the impact of visuality, maybe even too much.

自然光对于人类健康至关重要。尤其在北方，冬日的自然光照时间短，需要抓住所有可以利用的自然光。透明性成就了建筑的多维空间，是建筑必不可少的组成部分。

通过大玻璃窗辐射到室内的冷空气会让坐在附近的人感到不舒服，使用电热玻璃是非常实用的。薄而透明的导电金属薄膜夹在两层玻璃间通电供热。

相似的技术也适用于按下按钮就会变暗的玻璃。

钢化玻璃具有更好的抗冲击性。

安全玻璃是在两层玻璃之间夹一层塑料薄膜制成的。如果玻璃受到冲击，玻璃可能破碎，但塑料薄膜能将碎玻璃黏在一起。钢化玻璃则会碎成卵石状小颗粒。

氛围

建筑师彼得·卒姆托在他的《建筑氛围》一书中讨论了建筑氛围的重要性。氛围是建筑学研究和教学的核心议题之一，对环境进行规划时必须对其加以考虑（图 50）。

氛围是一种非常主观的现象，不同的人对环境的敏感性不同。

一些人关注于大的事件，而对于另一些人而言，小细节则传达了环境的意义。将环境氛围中的所有感受都纳入考虑范围是关键。对于不同的人而言，不同感官感受的意义是非常主观的。建筑师往往会过多地强调视觉效果的影响。

Zumthor has stated how everything in a building is of the atmosphere: smells, sounds and the echo in the space, for instance. How much can an architect design the atmosphere and affect it in advance? Some architecture publications have a skeptical view of this, and others state that an architect's most important task is to create a neutral space where users can create the atmosphere over time.

Many factors play a part in creating an atmosphere, or architectural feeling. The architect's own experiences are of the essence. Baselines learnt in books and photographs are secondary. Impressions of a certain moment, knowledge of history, interest in the hierarchy of different senses and many other little details are in the background of architectural experiences.

Jean–Paul Thibaud works at the Centre for Research on Sonic Space and the Urban Environment in Cresson. According to him, atmospheres are created naturally, like plants growing: effortlessly. Instead of swooning over spectacles, Thibaud raises people's ability to find different tones in the slow stream of everyday life, and he emphasises the importance of emotions in encountering the surroundings.

In a more brutal way, Gernot Böhme of the Institute for Practical Philosophy in Darmstadt, introduces the concept of atmosphere as part of the consumer society

彼得·卒姆托曾论述过建筑中切如何构成氛围，如空间中的气味、声音和回声。建筑师可以在多大程度上提前设计和影响建筑的氛围？一些建筑出版物对此持怀疑态度，还有一些出版物则表示建筑师最重要的一个任务是创造一个中性空间，让空间的使用者通过时间的累积来创造相应的氛围。

许多元素在创造氛围或者建筑感受时中扮演着重要角色，包括建筑师自身的经验。从书本和照片中学到的基本知识是次要的。对某一时刻的印象、对历史的了解、不同感官层次的兴趣以及其他许多小细节都是建筑体验的背景。

让–保罗·蒂博在克雷森的声音空间和城市环境研究中心工作。根据他的研究，如同植物生长一般，氛围是自然形成的，毫不费力。蒂博不是沉迷于表面景象，而是指出人们在缓慢的日常生活中发现不同风格的能力，并强调了情感在环境中的重要性。

达姆施塔特实践哲学研究所的格诺特·波默以一种更直接的方式将氛围概念作为消费社会的一部分，并从实用主义的角度定义了环境。罗

图 50　Spa in Vals, Architect Peter Zumthor　瓦尔斯温泉，建筑师彼得·卒姆托

and defining the environment pragmatically. Robert Venturi's observations from his book *Learning from Las Vegas* are an example of a postmodern stage set–like surroundings. Böhme also reminds of the long history of the concept of atmosphere.

Tonino Griffero (University of Rome Tor Vegata) tries to avoid an aesthetical Neo–Romantic interpretation of atmosphere. According to him, it is often assumed that trendy and manipulated environments are bad environments. More important than defining good or bad environments is seeing atmosphere as a political and ethical concept.

Giffero tries to fight against the over–psychologising of the world and thinking. People can balance their ability to understand their emotions and the surrounding world. The more you understand of atmospheres, the less the environment can manipulate you. This is the political and ethical benefit of understanding atmospheres.

In his texts, French anthropologist Marc Auge discusses places and non–places: even if an airport is a non–place for passengers, it is a place for a person working through their emotions and actions. The connection between atmosphere and emotions is a multidimensional question. Embodiment is also an important factor in understanding the atmosphere.

Herman Schmitz defines that the emotions in the atmosphere come to the people from the surrounding, not the other way around. People can sometimes create atmospheres but usually emotions are encountered in an environment involuntarily and you cannot affect them yourself.

伯特·文丘里在《向拉斯维加斯学习》一书中的观察是类似后现代主义舞台布景环境的一个案例。波默也提醒了我们氛围概念的悠久历史。

罗马托维加大学的托尼诺·格里费罗试图避免氛围的新浪漫主义的审美阐释。根据他的说法，人们通常认为流行的和被操控的环境是糟糕的环境。比定义环境的好坏更为重要的是将环境氛围视为一个政治和伦理概念。

格里费罗试图对抗世界和思想的过度心理化。人们能够平衡他们理解自身情绪和周围环境的能力。你对越理解氛围，环境就越不能操纵你。这是理解氛围的政治和伦理益处。

法国人类学家马克·奥格在他的文章中讨论了场所和非场所：机场对于乘客来说是一个非场所，但它对投入身心在此工作的人而言是场所。氛围和情绪之间的关系是一个多维问题而具象化也是理解氛围的一个重要因素。

赫尔曼·施米茨定义说氛围中的情绪来自周围的人。人们有时可以创造氛围，但是通常来说情绪是在一个环境中无意中出现的，你自己并不能影响情绪。

Postscript

To design healing architecture means that we should design for all our senses, particularly in the natural way so that most important senses come first. In architecture vision and touch are more dominating senses. Neuroscience will give some advice how people are responding to their environment and how we can design better environment for people.

Healing architecture is a new paradigm in architecture. We should not forget the artistic goals for architecture. To know more how our senses are working alone and with other senses can give us more knowledge how to design healing architecture.

Healing architecture with sustainable design can give us new perspectives designing better environment for human beings.

Buckminster Fuller always asked his colleagues, how much your building weighs? Nowadays architectural critics argue, is your building beautiful or not. I'd like to ask architects, is your building healing or not.

设计治愈性建筑意味着我们在设计时应该考虑我们所有的感官，尤其应采用顺应自然的方式，将最重要的感官感受放在第一位。在建筑中，视觉和触觉是更为主要的感官。神经科学将提供一些建议，告诉人们如何对环境做出反应，以及我们如何为人们设计更好的环境。

治愈性建筑是一种新的建筑范式，我们也不应该忘记建筑的艺术性目标。更多地了解我们的感官如何单独工作以及各个感官如何协同工作，能给我们更多的关于如何设计治愈性建筑的知识。

可持续的治愈性建筑能为我们提供新的视角，从而为人类设计更好的环境（图 51）。

巴克敏斯特·富勒一直在问他的同事，你的建筑有多重？今天，建筑评论家会讨论你的建筑美不美。我想问建筑师的是：你的建筑是不是治愈性建筑？

图 51 United Nationssustainable development goals 联合国可持续发展目标

图 52　Helsinki Railway Station platform roofing, Architect Esa Piironen, 2001
赫尔辛基火车站顶棚，埃萨·皮罗宁，2001

References / 参考文献

Alexander Christopher: The Nature of Order, 2002.

Ampuja Outi: Hyvä hiljaisuus, 2017.

Angelou Maya: Even the Stars Look Lonesome, Bantam Books, New York, 1998.

Arbib Michael: Why should Architects Care about Neuroscience in Architecture and Neuroscience, TWRB, 2013.

Coates Gary J. and Siepl–Coates Susanne: New Design Technologies: Healing Architecture, A Case Study of Vidarkliniken, 1998.

Cole John: Health Buildings: Insights on the Procurement Process in Care&Cure, 2016.

Heikinheimo Marianna: Paimio Sanatorium, Aalto University, 2015.

Jencks Charles: Can Architecture Affect Your Health, ArtEZ Press, 2012.

Kjisik Hennu: The Power of Architecture, Towards better Hospital Buildings, TKK, 2009.

Neutra Richard: Survival through Design, 1954.

Piironen Esa: Architecture and Materials China Electric Power Press, Beijing, 2021.

Piironen Esa: Measurement of Human Responses in Environment, TKK, 1978.

Tikkanen Kauko: A Study of School Children's Opinions in Classrooms with and without Windows, TKK A–os, 1981.

Ulrich Roger: View through a window may influence 3 recovery from surgery, Science 224, 1984.

Valtaoja Esko: Kohti ikuisuutta, 2017.

Wagenaar Cor: A Hospital Revolution in the Making in Cure&Care, 2016.

Images Indes / 图片索引

All images are from the archives of Esa Piironen Architects. 所有图片均来自建筑师埃萨·皮罗宁

译后记

张亚萍

在我之前翻译芬兰著名建筑师埃萨·皮罗宁的著作《建筑与材料》之初，听说埃萨先生还有一部新作品《治愈性建筑》即将完稿，我主动请缨，希望继续翻译这本《治愈性建筑》。为何我会对翻译《治愈性建筑》感兴趣？首先是翻译《建筑与材料》让我意识到埃萨先生是一位治学严谨、理论与实践皆强的建筑设计师，他的新作品必是值得一读。更重要的是，早年我读研究生期间的研究兴趣主要在针对特殊人群的景观空间设计上。最初我研究老龄化带来的针对老年人的景观设计，之后逐步关注针对儿童的城市开放空间设计。这些特殊的景观类型中，"治愈"和"友好"是常常提到的关键词。在华盛顿大学访学期间，我一直关注景观系丹尼尔教授的治愈性景观设计研究和实践工作。治愈性建筑与治愈性景观是相互关联的两个设计专项。在阅读了不少治愈性景观设计相关资料，以及做过"老龄化对城市社区景观设计的影响"等相关课题研究后，有治愈性建筑相关的建筑师著作我可以第一时间阅读、研究和翻译，何其开心。

治愈性建筑——像最舒适的家一样能治愈人们的建筑，早在 5000 年前人类在建造建筑时就有了相关的思考和实践。从古希腊的埃皮道罗斯遗址，到中世纪的第一家医院，再到现代芬兰建筑大师阿尔托的派米奥结核病疗养院，埃萨给我们呈现了以医院为主的治愈建筑的发展脉络。埃萨以建筑师的敏锐眼光，在环境心理学、神经系统学，乃至颇有争议的人智学等相关学科研究基础上，总结提炼出治愈性建筑设计的设计原则。而书的最后一部分，也是埃萨作为建筑设计师最为擅长的一部分经验总结，他娓娓道来治愈性建筑中各项设计元素的设计研究和设计运用，务实而极具设计参考价值。

2019 年底，我有幸与埃萨一起在苏州完成了一个实验性的教学工作坊。埃萨思维活跃、精力充沛，作为建筑设计实践经验极其丰富的芬兰第四代建筑师，其理论与实践均有多年的积累。短短几天的教学，令参与工作坊的师生受益匪浅。我深深佩服这位依然活跃在设计一线的充满活力、心态年轻的老人。在后期的居家翻译过程中，埃萨先生和我互通消息，就一些细节和具体内容又做了不少调整和修改。埃萨先生严谨的治学态度，令我印象深刻。我坚信这本基于多年设计实践和广泛理论阅读写出的精炼、实用、极具指导意义的专业著作《治愈性建筑》，绝对值得大家一读。

感谢我的朋友姚鹏对于医学专用词汇的解释与翻译建议。本书的翻译得到了电力出版社的大力支持与帮助，一并致谢在此书出版过程中所有帮助过我的人！

限于我的知识素养与翻译水平，翻译中难免有所不当之处，敬请读者斧正！